the farmers' market
guide to fruit

the
farmers' market
guide to fruit

jenni fleetwood

photography by lucy mason

SOURCEBOOKS, INC.®
NAPERVILLE, ILLINOIS

Series editor: Ljiljana Ortolja-Baird
Editor: Elizabeth Carr
Series designer: Bet Ayer
Designer: Bet Ayer
All photographs, except pages 36 & 65: Lucy Mason
Photograph on page 36: © Graham Parish/The Anthony Blake Photo Library
Photograph on page 65: © Maximilian Stock Ltd./The Anthony Blake Photo Library

Published by Sourcebooks
P.O. Box 4410, Naperville, Illinois 60567–4410
(630) 961–3900
FAX: (630) 961–2168

ISBN: 1–57071–632–3

Printed and bound in Spain

MQ 10 9 8 7 6 5 4 3 2 1

Author's dedication
This book is for my husband Bill, commis chef, chief taster, and constant supporter.

Author's Acknowledgments
Special thanks are due to Carole Young, the best cook I know, for contributing some
of her favorite family dishes, and for testing many of the recipes. Thanks, too, to
Judy Kelsey, for generously sharing the secret of her perfect pavlova.
For help in tracking down unusual fruits for testing, I am very grateful to Ben Mills of
Somerfield Food Markets Ltd, and for coming to the rescue over red currants, to
Chris Oaten of Rosemary & Thyme.

contents

introduction

What is it that comes in a glorious array of colors and shapes, is often exquisitely scented, tastes wonderful, and is very good for us? The answer, of course, is fruit. Nowhere else is nature so prodigal as in the provision of these fabulous foods, which, with vegetables, offer us so much in terms of nutrition. Most fruits are low in calories and contain only negligible amounts of fat. The sugar they contain provides valuable slow-release energy. They give us both soluble and insoluble fiber, to help regulate our digestive systems and control our cholesterol levels. A good source of vitamins, especially the antioxidant vitamins C and E, fruit also supplies us with a wonderful pharmacopia of phytochemicals. These remarkable plant compounds, some of which govern the color and flavor of fruit, are believed to protect against—or moderate the effects of—a wide range of diseases, including coronary heart disease, arthritis, osteoporosis, diabetes, hypertension, and even some cancers.

It's a small wonder that nutritionists recommend we eat at least five servings of fruit and vegetables every day. An average serving might be a large slice of melon or pineapple, one apple, orange, or banana, two plums or kiwi fruit, or a cupful (about 4 ounces) of berries. Fruit can be raw or cooked, juiced, frozen, canned, or dried. For maximum nutritional value, eat fruit raw. Try to avoid peeling it, as most of the nutrients lie just underneath the skin. It is important to eat a wide range of fruit, as each contributes a different balance of nutrients.

Eating more fruit is the easiest, most natural thing in the world, especially if you have it growing in your own garden. Alternatively, you may be able to pick your own from a nearby farm or—if you are really lucky—harvest it from the wild. The fresher the fruit the better, as some nutrients, such as vitamin C, decline quickly after picking.

When buying from a supermarket, choose one with a rapid turnover, and take a critical look at the produce. There should be a good range, including

exciting exotics, and the fruit should look fresh, bright, and inviting. If you consistently find fruit that is bruised or moldy, shop elsewhere. Select organic fruit when possible. It may not look as picture perfect as nonorganic produce, but the flavor will often be superior, and you can be sure it will have been grown without the aid of pesticides or other chemicals.

If you intend to use the skin on citrus fruit such as oranges, lemons, or limes, either by paring or grating the zest, it is important to choose unwaxed fruits. Buy fruit loose, when possible, so that you can see precisely what you are getting, and where berries are boxed, look underneath to make sure the bottom ones aren't squashed. Some types of fruit can be ripened successfully at home, provided they are not too underripe to begin with. Advice on selecting, storing, preparing, and cooking is given for all the fruits featured in this book, together with some of the most delicious recipes you could ever wish to find.

We've included refreshing fruit soups, perfect for starting or ending a meal, along with appetizers such as Fig and Goat's Cheese Bruschetta or Melon Pearls in Rosewater Dressing. Fruit features in both sweet and savory salads, with combinations such as pears with Roquefort cheese, strawberries and watercress, and watermelon slices with white peaches. Other savory dishes include Duck Breasts with Cherry Sauce, Beef & Kiwi Fruit Stir Fry, and Goujons of Salmon with Plum & Coriander Sauce. Desserts range from family favorites like Grandmother's Apple Pie and Blackberry Brown Betty to sophisticated treats such as Passion Fruit Pavlova, White Chocolate & Mango Terrine, Blueberry Muffins, and Lemon & Lime Love Cake. With your coffee, try Cape Gooseberries Dipped in Chocolate or delectable Stuffed Dates. Jams and jellies will help you store up summer's splendor, and there's even a selection of superb smoothies.

Five servings of fruit a day simply won't be enough!

apples

Apples are the world's favorite fruit, and have been ever since Eve took that fateful first bite. There is evidence to suggest that Stone Age man was partial to the odd apple, and apples were among the first fruits cultivated when man ceased to be a hunter-gatherer and settled in the fertile valleys of Asia Minor over 3,000 years ago. Apples were popular fruit in ancient Greece and Rome, and it was the Romans who introduced them to England.

Those early English apples were quite sour, however, and in medieval times children and nursing mothers were told to avoid them. In the monasteries, apples were valued for their laxative properties, and monks were advised to eat ten a day during Lent (apples were smaller then!). Perhaps this was one reason why new and sweeter varieties of apple began to be developed in monastery gardens, and by the time Henry VIII ascended the throne, orchards were a familiar part of the landscape.

johnny appleseed

When the Pilgrims sailed for America, they took apple seeds with them. The fruit flourished in its new environment, partly because of the efforts of John Chapman, better known as "Johnny Appleseed," who thought so highly of the fruit that he spent forty-nine years planting apple seeds as he walked through Pennsylvania, Ohio, Kentucky, Illinois, and Indiana.

Today there are over 7,000 varieties of apples worldwide, ranging from tiny crab apples (the ones the monks were made to eat) to the massive Howgate Wonders, which weigh in at around 3lb. In England, apples are classified as eaters or cookers, with Cox's Orange Pippin the best known of the former, and Bramley's Seedling taking the honors in the second category. In America, as elsewhere, the boundaries between cookers and eaters have become blurred, and many apples are all-purpose varieties. Favorites include Jonathan, Paula Red, and Northern Spy (all good for pies) and the crisp, tart Spartan or the sweet, crunchy Macoun, which also make good eating. For salads, the hard, sweet Red Delicious is a good choice, as is the Cortland, which is slow to brown. Idared and McIntosh are ideal for applesauce.

granny smith

Apples are cultivated in the temperate regions of the southern hemisphere as well as the northern, and one of the world's most famous apples comes from Australia. This is the Granny Smith, which was developed in a suburb of Sydney in 1868 by Mrs. Thomas Smith. Bright green and shiny, the Granny Smith is exceptionally juicy, and has a sweet yet tangy flavor.

New Zealand introduced the world to the Gala apple, a butter yellow fruit striped with pink. When very fresh, the flesh is beautifully crisp and juicy. Galas are becoming increasingly popular in Europe and America, thanks to their sweet and mildly aromatic taste and dense, crisp flesh.

good travelers

Many varieties of apple travel well, and at any time your local supermarket is likely to stock several different types. Colors range from creamy yellow through to deep crimson, with almost as much variation in the flesh beneath. This can be anything from pure, almost greenish white to a buttery yellow. Texture is also variable. Some apples are crisp and crunchy, others quite soft. Some apples taste so sharp that they set your teeth on edge, while others are sweet and honeyed. You can find apples that taste faintly of pineapple, banana or even aniseed.

fresh from the tree

Although apples, by and large, travel well, the very best fruits are usually those that have been freshly picked. Some of the most delicious apples are not cultivated commercially, but can still be found in gardens. If you grow your own, you are lucky indeed. If not, the next best thing is to investigate your local farmers' market, or look out for fruit farms with roadside stalls.

an apple a day

The old adage is only partly right. Apples are not particularly nutritious, but do contain vitamin A and C, calcium and phosphorous, as well as significant amounts of potassium. An average apple provides nearly 4 grams of dietary fiber and most of that is cholesterol-lowering soluble fiber. The complex carbohydrates in apples provide plenty of slow-release energy.

selection and storage

If possible, buy apples loose. When apples are bagged it is difficult to see what you are getting, and you certainly can't sniff the fruit to determine its fragrance. The apples should be firm and free of blemishes and bruises, with taut, not wrinkled, skin. The first thing most of us do when we buy apples is to polish them and put them in a fruit bowl (usually artistically arranged where they will be most

admired). This is fine if you are going to eat the apples immediately, but most types of apple keep much better in the refrigerator. Put them in the salad crisper, preferably in a plastic bag with ventilation holes, and wash them just before use. For long term storage, wrap apples in paper and place them in a single layer on a shelf in a cool place, such as a garden shed. Check them frequently, remembering that "one rotten apple rots the whole bushel," and make sure the mice don't get them before you do.

preparation

Everyone eats their apples differently. The world is divided into chompers, who just sink in their teeth and crunch noisily, and nibblers, who peel the apples and cut them into neat pieces, which they convey delicately to their mouths. Some people always eat the skin; others insist on peeling the fruit, often in a continuous spiral. It used to be said that if you peeled an apple that way, without breaking the spiral, and threw it over your left shoulder, the initial it formed would be the first letter of your lover's name.

If you intend to cook the apples, check the recipe for guidance on preparation. For baking whole, remove the core and seeds with an apple corer (a metal tube pointed at one end, which is driven into the center of the fruit from the top, then pulled out to leave a neat hollow). A sharp knife can be used if you do not have a corer. If you leave the skin on, slit it all the way around the apple about halfway down;

this will stop it from bursting. Alternatively, peel the top half of the skin, then coat the apple with butter and flavored crumbs. If the recipe calls for sliced apples, the easiest way to do this is to quarter the apple, then use a vegetable peeler to remove the skin thinly from each piece. Cut away the core with a sharp knife. Slice each wedge lengthwise or widthwise, depending on whether you want small or large slices.

Apple flesh discolors very rapidly, so whether you are slicing or chopping the fruit, put it into acidulated water straightaway. This is water to which a little lemon juice or vinegar has been added. Raw apple, grated and mixed with a little honey, is a marvelous pick-me-up. Apple juice is a refreshing drink, and apples are also used to make cider.

cooking

Apples are very versatile. They can be baked whole in the oven or microwave, the centers filled with dried fruit; they can be sliced and baked in tarts or pies; they can be cooked to a purée for applesauce, to be served solo or used in desserts. Apples are also good in savory dishes, particularly with pork or goose, and are traditionally cooked with red cabbage. They contain plenty of pectin, so jams and jellies that contain apples will set well.

fresh apple muesli

Peel and core two eating apples and chop the flesh roughly. Put it in a blender with the juice of 1 lemon, 1/4 cup pecan nuts, 1 tablespoon sugar, and 2/3 cup half-and-half. Whizz until smooth, then spoon into a bowl and stir in 1/4 cup rolled oats. Cover and leave overnight in the refrigerator. Next day, stir in 1 tablespoon honey and 1/3 cup raisins. Serves 1–2.

grandma's apple pie

With its domed top, tender pastry, and delicious, spicy filling, this is a real winner. Enjoy it hot or cold, with whipped cream or ice cream.

Serves 8

3 cups all-purpose flour
½ teaspoon salt
6 tablespoons butter
6 tablespoons lard or vegetable shortening
milk, to glaze
sugar, for dredging

Filling

2 pounds cooking apples
2 teaspoons lemon juice
6 tablespoons raisins
1 teaspoon pared orange zest
1 cup light brown sugar
4 teaspoons cornstarch
1 teaspoon ground cinnamon
1 teaspoon freshly grated nutmeg
4 teaspoons butter

Preheat the oven to 400°F. Make the pastry by mixing the flour and salt in a bowl, rubbing in the butter and lard or shortening, then adding just enough ice water to bind.

Make the filling. Quarter, peel, and core the apples, then slice them thinly into a bowl. Toss with the lemon juice, then add all the remaining filling ingredients, except the butter, and mix lightly.

Roll out just less than half the pastry on a lightly floured surface and line a 10-inch pie plate. Add the filling, mounding it in the center, then dot with the butter.

Roll out the remaining pastry and make a lid for the pie. Seal and crimp the edges and cut a couple of steam vents. Brush the surface lightly with milk and sprinkle with sugar. Bake for 15 minutes, then lower the oven temperature to 350°F and bake for 20 minutes more. Serve hot or cold.

apples

12

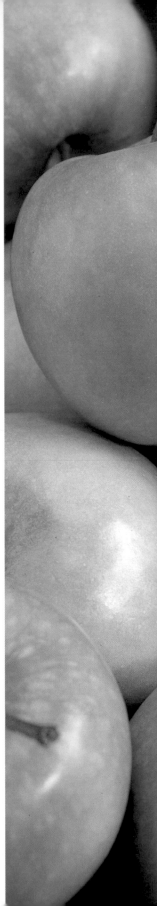

Granny Smith apples

baked apples with walnut jackets

The contrast between the nutty, sweet crumb crust, and the juicy and slightly tart apple beneath makes these absolutely irresistible.

Serves 4

1 cup walnuts
4 gingersnaps, broken
⅔ cup light brown sugar
1 teaspoon ground cinnamon
½ teaspoon Chinese five spice powder
4 large tart apples
3 tablespoons butter, melted
plain yogurt, to serve

Preheat the oven to 350°F. Blend the walnuts, gingersnaps, brown sugar, cinnamon, and five spice powder in a blender or food processor until the mixture forms fine crumbs.

Core the apples, then score the skin on each one around the circumference, two-thirds of the way down the fruit. Peel off all the skin above the mark. Take a thin slice off the bottom of each apple, if necessary, so that they stand straight.

Put the apples in a baking dish and fill each one with the walnut mixture, packing it down firmly. Brush the exposed flesh on the top of each apple with the butter and press the remaining walnut mixture onto it so that each apple is topped with a walnut jacket. Drizzle any remaining butter over, being careful not to dislodge the crust.

Bake the apples for 45 minutes or until tender. Serve hot or cold with plain yogurt.

tart tatin

A French classic, this upside-down tart looks spectacular and is very easy to make if you use storebought puff pastry.

Serves 8

10 ounces puff pastry, thawed if frozen
cream or ice cream, to serve
Filling
6 eating apples, such as Golden Delicious
1 tablespoon lemon juice
6 tablespoons butter
6 tablespoons sugar

Slice the apples into quarters. Remove the core and peel from each quarter, then use a fork to score the rounded side. Cut each quarter in half widthwise, then toss the pieces in the lemon juice.

Melt the butter in a 9-inch heavy-bottomed, ovenproof frying pan. Stir in the sugar until it has melted, then remove the pan from the heat.

Arrange the apple quarters, scored side down, in concentric circles in the pan, packing them quite tightly together. Put the pan over a low heat and cook gently, without disturbing the apples, for about 15 minutes or until they have begun to caramelize.

Preheat the oven to 400°F.

Roll out the puff pastry on a lightly floured surface to a circle slightly larger than the top of the ovenproof frying pan. Wrap it over the rolling pin, then carefully place it on top of the apples

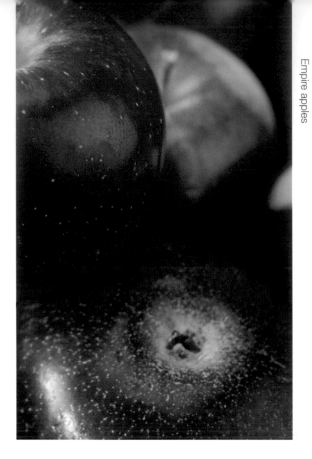

in the frying pan. Taking care not to burn your fingers, carefully tuck the edges inside the pan.

Transfer the frying pan to the oven and bake the tart for 20–25 minutes, until the pastry has risen well and is golden brown.

Let the tart cool for 5 minutes, then gently ease a knife between the top crust and the pan. Invert a plate on top, then carefully turn both pan and plate over together so that the apples are now on top.

Serve warm, with cream or ice cream.

variations
• Other types of fruit can be used to make similar tarts. Pears, plums, nectarines, and apricots work well, as does rhubarb.

apples

14

applesauce sundae

Layered in wine glasses, this looks very pretty, and the combination of smooth apple purée and crisp crumbs is a sure winner.

Serves 4

1½ pounds cooking apples
¼ cup water
¾ cup sugar
1 stick butter
3 cups fresh brown or rye breadcrumbs
8 tablespoons redcurrant jelly
½ cup heavy cream

Peel, core, and chop the apples. Put them in a heavy-bottomed pan with the water. Cover and cook over a moderate heat until the apples are very soft, stirring frequently. Purée them in a blender or food processor. Stir in 6 tablespoons of the sugar and leave until cold.

Melt the butter in a large frying pan and add the breadcrumbs and remaining sugar. Cook over low heat, shaking the pan frequently, until the crumbs have absorbed the butter and become crisp. Set the pan aside.

Warm the redcurrant jelly if necessary, so that it can be spooned. Layer the apple purée, jelly, and browned crumbs in four wine glasses, keeping back a little of the jelly for the decoration. Whip the cream until soft peaks form, then swirl it on top of each dessert. Drizzle the remaining redcurrant jelly over the cream and serve.

variation
• Use gooseberry purée instead of apple.

Empire apples

apples

15

pears

Beautifully scented and seductively shaped, pears have been popular for centuries. Related to the apple, they originally grew wild in parts of Europe, Central Asia, and Northern China. Little is known about where and when they were first cultivated, but the Romans grew at least 30 different varieties, and it was one of the many fruits that found favor with the ruling Medici family during the Renaissance in Italy, when it was appreciated as much for the way it could be used in lavish table settings as for its wonderful flavor.

France is famous for its pears, and many of the world's classic pear recipes come from that country. It was the French who first classified pears by shape: the oval *bézy*, the ball-shaped *bergamotte*, the quince-like *chretien*, the *calabash* (with an elongated neck), and the *colmar*, which looks like a spinning top.

The first pear tree was reputedly planted in America in 1620.

thousands of varieties

Pears are now grown in all the world's temperate regions, and there are more than 5,000 different varieties. In Britain, the most popular are Williams (widely regarded as having the finest flavor, but with poor keeping qualities), the slender, slightly granular Conference, and the large, juicy Comice. America is a major pear producer. The first fruits available, often as early as August, are the Red and Yellow Bartletts. Sweet, aromatic, and juicy, they are excellent for eating in the hand, but are also good for cooking and canning. These are followed by Anjou pears. There is a green variety, which stays the same color even when ripe, and a Red Anjou, which turns a deep maroon. Boscs are large, with thin necks. They are highly aromatic, but can be somewhat tart. As in England, Comice pears are valued for their sweet, juicy flesh, while for snacks, the smaller and very sweet Seckel and Forelle pears are perfect. Most of the pears we buy today are all-purpose varieties, suitable for both eating and cooking. True cooking pears are unpalatable raw, but are good for stewing. They are sometimes sold on farm stalls.

selection and storage

Pears are picked while they are still hard. By the time they arrive in our stores, they should have begun to ripen, and should yield slightly when lightly pressed at the stem end. At this stage they are perfect for cooking, but if you want to eat them fresh, let them ripen further at room temperature. Watch them closely though, as pears easily become overripe. Ripe pears can be kept in the refrigerator for a short time.

preparation

There is no need to peel pears if you are eating them fresh. Just halve or quarter them and cut out the cores. If you are cooking pears whole, peel them and remove the cores from underneath, using a small knife. The cavity can be stuffed with a crumb or dried fruit mixture. Unless you are cooking them right away in a mixture that contains some acid (such as wine), brush them with lemon juice to avoid discoloration.

cooking

Pears are delicious poached or baked. The classic way to cook them is in a red wine syrup, with spices, but they also taste superb with a butterscotch or maple sauce. Savory roast pears are also excellent; pop them into the roasting pan toward the end of cooking.

asian pears

There are several varieties of Asian pear. They tend to be round, not elongated, and the skin color ranges from bronze to russet. They are sometimes called Nashi (nashi means pear in Japanese) or apple pears, the latter reflecting not just their shape, but also the texture, which is crisp and juicy. Unlike European pears, which soften so that the flesh becomes almost buttery, Asian pears stay crisp after storage. They make excellent eating.

nutrition

In terms of nutrition, pears are quite a poor choice, yielding a small amount of vitamin C, and some potassium. They are a good source of dietary fiber.

fanned pears

An attractive way to present pears is to fan them: halve them and remove the cores, then make several parallel slices from stern to stem, stopping before you get to the top, so the pears remain intact at that point. Put them, cored side down, on a plate, and press down on them so that the sections fan. Brush the flesh with lemon juice to prevent discoloration.

pear and roquefort salad

Peppery arugula, honey-sweet pears, and salty Roquefort cheese combine beautifully in this salad, which would make an elegant first course.

Serves 6

2 Romaine lettuce hearts
6 ounces arugula, about 6 cups
6 small ripe pears

Dressing
⅓ cup Roquefort cheese
scant 1 cup crème fraiche or sour cream
salt and pepper

Make the dressing by blending the Roquefort cheese and crème fraiche or sour cream in a blender or food processor. Season with plenty of salt and pepper.

Separate the lettuce hearts into leaves. Wash and dry all the salad leaves. Fan three or four lettuce leaves on each of six salad plates. Put the arugula leaves in a bowl and add ¼ cup of the dressing. Toss well.

Cut the pears in half and carefully scoop out the cores with a teaspoon. Place each pear half in turn on a board, cut side down, and make several parallel cuts from the rounded side to the stem end, stopping just before you reach the stem so that each half remains intact at that point. Place two pear halves on each salad, fanning them out by pressing down on them gently with your hand, and arranging them over and between the lettuce leaves.

Spoon the dressed arugula on to the plate. Serve at once, with the extra dressing in a small pitcher.

tip
• If you do not intend to serve the salad immediately, brush the pear flesh with lemon juice to prevent discoloration.

Conference pears

roast lamb with pears

If you've never roasted pears, do yourself a favor and try this wonderful dish. The hazelnut and pear stuffing is moist and flavorsome, making this ideal for a special occasion.

Serves 6

1 cup hazelnuts
4 slices white bread, crusts removed
5 dessert pears
5 tablespoons butter
1 small onion, chopped
1 tablespoon chopped fresh sage
salt and pepper
4 pounds lamb shoulder, boned
1 tablespoon sunflower oil
1 teaspoon ground allspice
several strips orange zest, to garnish

Preheat the oven to 375°F. Spread out the hazelnuts on a baking sheet and roast for about 10 minutes or until golden brown, shaking the pan occasionally. Let them cool slightly, then rub off the skins in a clean kitchen towel.

Crumb the bread in a food processor, then tip it into a bowl. Chop the hazelnuts coarsely in the food processor and add them to the bowl.

Peel and core two of the pears and chop them finely. Melt two-thirds of the butter in a small pan and cook the onion until soft. Stir into the breadcrumb mixture with the chopped pears and sage. Season well.

Trim the boned lamb neatly and remove the excess fat. Use a rolling pin to flatten it slightly, then spread the stuffing over half the surface. Fold the rest of the lamb over to make a neat package and tie this with string.

Grease a roasting pan with the oil, add the meat and roast it for 1¼ hours.

Peel the remaining pears and cut them in half lengthwise. Place them in the roasting pan, rounded sides up, sprinkle with the allspice and dot with the remaining butter. Return the roasting pan to the oven and roast for 15 minutes more, or until the meat is tender and cooked through. Baste the pears occasionally with the pan juices.

Transfer the meat to a heated platter and surround it with the roasted pears. Garnish with the orange zest strips and serve.

tip

• Ask the butcher to give you the lamb bones and use them to make a thin gravy to serve with the roast. Put the bones in a pan and add 1 chopped carrot, 1 chopped onion and 1 tablespoon chopped sage. Pour in 2½ cups water and add plenty of salt and pepper. Bring to a gentle simmer, cover and cook for 20 minutes. When the lamb and roast pears are cooked, transfer them to a heated platter, cover with tented foil and leave to stand. Pour away the excess fat from the roasting pan, add the lamb stock and stir well, scraping in any sediment from the base of the pan. Boil the gravy until reduced to about 1¼ cups. Adjust the seasoning and serve with the meat.

pears

20

maple & pecan poached pears

This dessert tastes good hot, but is even better cold, when the sauce sets to a delicious coating.

Serves 4

1 lemon
1 cup water
8 tablespoons pure maple syrup
1 cinnamon stick
4 firm pears
¼ cup light brown sugar
2 tablespoons butter, diced
1¼ cups heavy cream
2 egg yolks
chopped pecan nuts, to decorate

Pare a long strip of zest from the lemon and put it in a large pan. Squeeze the lemon and set aside 1 tablespoon of the juice. Add the remaining juice to the pan, with the water and 2 tablespoons of the maple syrup. Add the cinnamon stick and heat, stirring gently all the time.

Peel the pears, leaving them whole. Add them to the syrup, carefully spoon it over them, then cover the pan. Poach the pears, basting them occasionally, until they are transparent and just tender, but still firm enough to hold their shape well. The timing will depend on the type of pears used, and their size, so check them frequently. When they are cooked, carefully transfer them to a dish and leave them to cool, spooning the syrup over from time to time.

When cool, place in the refrigerator and chill until ready to serve.

Meanwhile, put the brown sugar in a heavy-bottomed pan and add the remaining maple syrup, the reserved lemon juice, and the butter. Heat gently, stirring, until the mixture is smooth. Remove it from the heat.

Heat the cream in a separate pan. When it is on the verge of boiling, pour it into the brown sugar mixture in a steady stream, stirring all the time.

Beat the egg yolks with 6 tablespoons of the cream mixture. Stir the mixture back into the pan and heat gently, stirring all the time until the sauce starts to thicken. Pour it into a pitcher, cover the surface closely and let it cool completely.

Drain the pears, cut in half, and remove the cores. Fan them on dessert plates, spoon the maple sauce over, decorate with the chopped pecans and serve.

tip

• To obtain the maximum amount of juice from lemons, place them in hot water for 30 seconds, then drain them and roll them on a hard surface before squeezing.

Nashi pears

quinces

A fruit that deserves to be much more widely used, the quince comes from the same family as apples and pears, and can be shaped like either. The flavor has something in common with both, but is much more intense.

Quinces are believed to have originated in Persia and still grow wild in much of Iran and Iraq. In warmer climates the fruit tends to be soft and juicy, and can be eaten straight from the tree, unlike cold climate quinces, that have hard, unpalatable skins, often covered in a downy jacket, and must be cooked.

One of the oldest cultivated fruits, quinces were well known to the Ancient Greeks and Romans. Pliny wrote about the fruit's medicinal properties, and its efficacy in warding off the evil eye. The botanical name—cydonia—refers to Cydon in Crete, where particularly fine quinces grew. These were glorious golden fruit, the "golden apples" mentioned in the legends and myths of Greece and Rome. Remember the story of Paris, who was ordered by Zeus to award a golden apple to the goddess judged to be the most beautiful? Paris chose

Aphrodite, goddess of love, which is singularly appropriate for a fruit that has always been associated with love, and that is regarded as an aphrodisiac in western Asia.

fragrant fruit

The most notable characteristic of quince is its wonderful fragrance. A single fruit can perfume a room, and in medieval times quinces were often used for this purpose. Don't relegate them solely to the role of room freshener, however. Most quinces may be unsuitable for eating raw, but when cooked, the grainy, rather unprepossessing flesh becomes beautifully tender and flavorful as it turns a delicate rose or apricot color. A good source of vitamin C, quinces are also high in pectin, and make wonderful jelly and jam. It was the quince, not the orange, that was used to make the first marmalade (the Spanish word for quince is marmelo). In Spain, where the fruit has been popular for centuries, it is made into a delectable paste, called membrillo.

japonicas

There are several varieties of quince, including the japonica, which is often grown as an ornamental shrub. The fruit of the japonica is less aromatic, but still makes a very good jelly.

selection and storage

Quinces are seldom sold in supermarkets, but can be bought in the fall from farmers' markets or roadside stalls. Apple-shaped quinces are regarded as having the best flavor, but all quinces are worth cooking. Unless you live in the tropics, the fruit you buy will probably be very hard and feel surprisingly heavy. As compensation for their tough skins, most quinces keep exceptionally well.

preparation

The hardness of the common quince makes it quite challenging to prepare. If slicing it proves a struggle, poach the whole fruit for 10 minutes, then quarter it as you would an apple, and remove the core and small brown seeds. If you are making quince jelly, grate or grind up the entire fruit—peel, flesh, and seeds—in a food processor.

cooking

Quinces can be cooked in much the same way as apples. Iranians use them in savory dishes, like the famous *dolmeh beh*, in which the fruits are stuffed with a spicy beef and split pea mixture. Quinces also make wonderful ice creams and sorbets, and have a particular affinity for apple. A mixture of quince and apple makes an excellent pie filling.

quince jelly

This has the most beautiful rose or apricot color, depending on the type of quince you use. It tastes superb with roast game birds, venison, or lamb.

Makes about 4 8-ounce jars

1 pound quinces
4 cups water
juice of ½ lemon
sugar (see method)

Remove the stalks from the quinces, but do not peel or core them. Cut them into large pieces and then grate them, including the skin, seeds and all, in a food processor.

Put the quinces in a preserving pan and pour in the water and lemon juice. Bring to a boil, lower the heat, cover and simmer for 30–45 minutes, or until the quince flesh is soft. Pour the contents of the pan into a jelly bag suspended over a large bowl. Leave overnight to drain.

The next day measure the liquid in the bowl, then pour it into a clean preserving pan. Add 1 cup sugar for every 1¼ cups of liquid. Stir over low heat until the sugar has dissolved. Bring the mixture to a boil and boil rapidly until the setting point is reached. If you have a sugar thermometer, it should read 220°F. Alternatively, test by spooning a little of the jelly onto a saucer that has been chilled in the refrigerator. Leave for a minute, then gently push the jelly with your fingertip. It should wrinkle.

Ladle the jelly into hot jars. Seal with canning lids and process for 5 minutes in a boiling-water bath.

quince

23

plums

Few fruits are as prolific as plums. There are thousands of varieties, and as many ripen at different times; it is rare to fail to find them at the market. As with most fruits, it is difficult to pinpoint where plums originated, but they are known to have grown wild in Asia at least 2,000 years ago. They were introduced into Britain during the 15th century, and rapidly became a very important fruit. Plums were among the earliest cultivated trees, and "plumb pudding"—the forerunner of the famous British Christmas pudding—was served as a special treat for all sorts of celebrations, not just Christmas. Prunes, which are dried purple plums, were used in plum pudding until the 16th and 17th centuries, but after that time, raisins became the more popular ingredient.

Plums come in a wide array of skin colors, from green (gages) to red, purple, and even black. The flesh can be yellow, orange, or red and the flavor ranges from tart to mild to sweet. Some plums are quite firm and crunchy, others tender and juicy. Most plums are oval, but some are round, and size also varies markedly. European plums are fairly small and tart, with blue, green, or black skins. The pits are usually free, and come away cleanly when the fruit is cut in half. Japanese plums are larger and lusher, with scarlet, yellow, or magenta skins. The flesh tends to cling to the pits. Some plums must be cooked; others can be eaten fresh. Early Laxton and Victoria plums are popular British varieties, while in America the vermilion Burbank and the red-purple Santa Rosa are among the best known dessert plums.

nutrition
Plums are a good source of vitamin A and contain some vitamin C. They also contain antioxidants.

beach plums
Some varieties of plum are too bitter to eat raw, but are delicious cooked. Bullaces and Damsons fall into this category, as do the American beach plums, cherry-size fruit that are peculiar to New England coastal scrub. They have a tart flavor, so are seldom

eaten fresh, but can be transformed into wonderful jelly or jam. Sloes are extremely bitter plums, about the size of olives. They are much too bitter to eat raw, but marry them with gin and they make a superb drink (sloe gin).

selection and storage

Whichever type of plum you buy, look for plump, blemish-free fruit that yields when gently pressed. Avoid any plums that look dry or shrivelled, or that have a "musty" smell. Store ripe plums briefly in a perforated plastic bag in the refrigerator, but use them as soon as possible after purchase, as they rapidly become overripe.

preparation

Ripe eating plums need no special preparation, other than washing, and are delicious simply eaten in the hand. For cooking, plums can be baked whole (in which case you should make a small slit in the skin of each fruit to prevent it from bursting). Alternatively, cut the plums in half, following the natural seam, then twist the halves apart and ease out the pits with the tip of a sharp knife.

cooking

One of the best ways to eat plums is also the simplest. Serve them as a compote. They can either be stewed in a little water and sugar on top of the stove, or baked in a low oven. Baking is preferable, as the fruit is less likely to disintegrate. Plums are also delicious in pies and crumbles. Don't simply use plums in sweet dishes—take a tip from Asian cooks and make a savory plum sauce to serve with baked ham, pork chops, or fried fish. Try pickled plums and plum chutney, too.

goujons of salmon with plum & coriander sauce

Strips of salmon, deep-fried and served with a tart and spicy plum sauce, make a wonderful appetizer. The sauce needs to be very cold, so make it well ahead of time.

Serves 4

1 pound tail end of salmon, skinned
½ cup all-purpose flour
2 eggs, beaten
oil for deep frying
watercress or arugula leaves to garnish

Plum and Coriander Sauce
2 tablespoons butter
1 small onion, finely chopped
2 garlic cloves, crushed
1½ teaspoons ground coriander
¾ pound plums, halved and pitted
1 cup water
1 tablespoon balsamic vinegar
salt and pepper

Start by making the sauce. Melt the butter in a small pan, and fry the onion and garlic for 5 minutes or until the onion has softened.

Stir in the coriander and cook for 1 minute, then add the plums and water. Simmer for about 15 minutes, or until the plums are very soft.

Purée the mixture in a blender or food processor, add the balsamic vinegar and pulse briefly to mix.

Scrape the mixture into a bowl, season it generously and leave it to cool. Chill thoroughly.

Cut the fish into short strips, 1 inch wide and 3 inches long. Season the flour with a little salt and pepper and spread it out in a shallow bowl. Pour the beaten eggs into a separate bowl. Heat the oil for deep frying.

Dip a few fish strips at a time into the seasoned flour, coating them completely, then dip each in turn in egg, letting the excess drip off, and fry in the hot oil. When the strips (goujons) are golden, after about 2 minutes, remove them with a slotted spoon, drain on kitchen paper towels, and keep hot while cooking the rest of the goujons. Place the goujons on a platter, garnish with the watercress or arugula, and serve with the chilled plum sauce.

fresh plum compote with cranberry cheese dressing

Poached plums and almonds make a delicious dessert, especially when served with a creamy, pale pink sauce flavored with cranberry jelly.

Serves 4

1 ½ pounds plums
1 cinnamon stick
4 tablespoons whole blanched almonds
½ cup sugar
3 tablespoons water
Cream Cheese Dressing
½ cup cream cheese
1 tablespoon lemon juice
⅔ cup heavy cream or crème fraiche
2 tablespoons cranberry jelly

Preheat the oven to 300°F. Make a small slit in the skin on each plum and then spread them out in a baking dish. Add the cinnamon stick, then add the blanched almonds, pushing them down among the fruit so they are evenly distributed. Sprinkle the sugar over the top. Drizzle with the water and bake for 45 minutes to 1 hour, stirring occasionally, until the plums are soft but have not collapsed.

Meanwhile, make the dressing. Beat the cream cheese until softened, gradually working in the lemon juice and cream or crème fraiche.

Remove the cinnamon stick from the plums and transfer them to a serving dish. Stir the cranberry jelly into the cream cheese dressing to sweeten it slightly and add a blush of color.

Serve the plums warm or at room temperature, with the dressing.

tips
• The cooking time will vary slightly depending on the size and variety of the plums that you use; it is a good idea to check them frequently during the baking time, to make sure they do not become too soft.
• Use either quince jelly or redcurrant jelly as an alternative to cranberry jelly for flavoring and coloring the cream cheese dressing, if you prefer.

plums

27

peaches & nectarines

What could be more perfect than a ripe peach picked from the tree on a sunny evening? The warmth of that downy skin; the sudden sweetness as you take your first bite; the tender flush of the flesh? Peaches provoke passion, and it isn't surprising that poets and painters have drawn inspiration from them. Nectarines come from the same family, but are much more restrained. Subtle and smooth-skinned, they sit on the sidelines, but are just as tasty as their more flamboyant relatives.

Peaches originated in China, but were soon embraced by the Persians, who served them as salads and sherbets, poached them with rosewater and cardamoms, and even incorporated them in savory dishes.

Today, peaches are grown in many parts of the world, including Europe, South Africa, Chile, and Australia. America has been growing peaches commercially since the beginning of the 19th century, and now produces a quarter of the world's total supply of fresh fruit. Depending on how enthusiastically the flesh adheres to the stone, peaches are categorized as clingstone, semi-freestone, and freestone. American varieties include Regal, Goldprince, Bounty, Flame Prince, and Late Gold. White peaches are regarded as being superior in flavor to varieties with yellow flesh, but this is largely a matter of taste. Nectarines have white, yellow, or pink flesh. They are also widely cultivated.

nutrition
Both peaches and nectarines contain useful amounts of vitamins A and C.

selection and storage
Look for peaches and nectarines that have been ripened on the tree. Fruit that has been picked too early will soften when kept at room temperature, but it will grow no sweeter. Avoid any bruised or damaged fruit, and handle with care after purchase, placing it at the top of your bag. If you are going to eat the fruit within a day or so, it can be kept in a fruit bowl. Alternatively, wrap it and store it in the refrigerator for up to five days.

preparing

If the fuzz on peaches offends you, rub it off under cold running water. From a healthy eating perspective, it is best to eat the skin as it provides fiber and nutrients, but you can peel peaches if you prefer. Nick the skins, then plunge the fruits into boiling water for a few seconds. Lift them out using a slotted spoon, and dunk them in cold water. The skins should slip off easily. Nectarines do not need to be peeled. Cut the fruit in half, if you like, and remove the pit. If it comes away easily, you've chosen a freestone variety; if it sticks, slice the half holding the pit in half again, so that you can access the pit, then cut around it. Peach and nectarine slices will discolor if left to stand for a long time; to prevent this, brush them with lemon juice.

cooking

Peaches and nectarines can be cooked in precisely the same way. A delicious way of serving them is to cut them in half, fill the centers with almond paste or a mixture of macaroon crumbs moistened with Amaretto liqueur, then grill or bake them. Peaches and ham or bacon are a good combination, and, perhaps surprisingly, both peaches and nectarines go very well with shrimp. Peach upside-down cake is an old favorite. Both fruits make marvelous pies and tarts, and can be used for jam, pickles, or chutneys.

peaches in champagne

Slice peaches or nectarines into champagne flutes. Sprinkle about two teaspoons of sugar into each glass and add a splash of peach brandy if you have any. Finally, pour in champagne or sparkling white wine and serve.

shrimp & peach brochettes

Brochettes can sometimes be a little dry, but marinating the shrimp and adding juicy peach pieces prevents that from happening here, and the flavors complement each other beautifully.

Serves 6

½ cup extra virgin olive oil
2 tablespoons fresh lime juice
2 garlic cloves, crushed
1 tablespoon chopped fresh dill
1½ pounds raw large shrimp, in the shell
2 peaches, halved, pitted, and cut into
    large chunks
24 bay leaves
fresh dill sprigs and lime wedges, for serving

Mix the olive oil, lime juice, garlic, and dill in a bowl. Peel the shrimp, leaving the tails intact, and add them to the bowl. Toss to coat, then set aside for 20 minutes.

Preheat the grill. Drain the shrimp, reserving the marinade, and thread them on 12 metal skewers, alternating with the peach chunks. Slip a bay leaf on both ends of each skewer.

Grill close to the heat, turning once and brushing the shrimp occasionally with the reserved marinade, for 4–5 minutes or until the shrimp have turned pink and are cooked through. Transfer the brochettes to individual plates, garnish with the dill sprigs, and serve with the lime wedges.

pan-fried panettone with nectarines

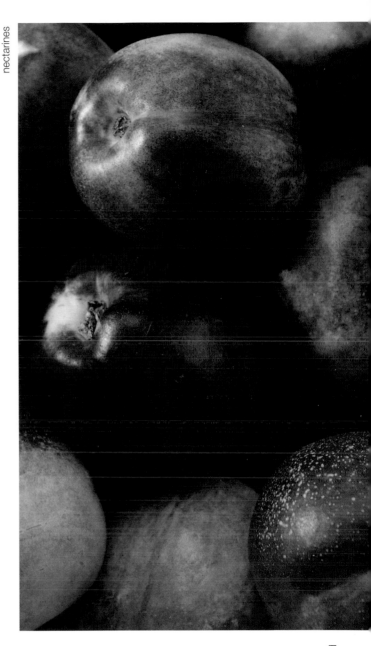

Panettone makes marvelous French toast, but if you can't locate it, use brioche or any fruited bread instead.

Serves 4

2 eggs
2 tablespoons sugar
½ cup milk
½ cup half-and-half cream
¼ teaspoon pure vanilla extract
4 tablespoons butter
2 nectarines, sliced
2 tablespoons light brown sugar
4 slices panettone (or 2 large slices, halved diagonally)

Put the eggs in a shallow bowl large enough to hold one of the pieces of panettone. Stir in the sugar, then the milk, cream, and vanilla extract.

Melt half the butter in a large frying pan, add the nectarines and sprinkle them with the brown sugar. Cook, shaking the pan frequently, until soft and caramelized. Spoon into a dish and keep hot.

Melt the remaining butter in the pan. Add a slice of panettone to the egg mixture, turning to coat it on both sides. Transfer the panettone to the pan and fry it over medium heat for about 2 minutes, then turn it over carefully and cook the other side for 2 minutes. If your pan is large enough, cook 2 slices at once. Prepare the remaining panettone in the same way, adding more butter to the pan if necessary, and keeping the cooked panettone hot. Serve the panettone with the nectarines.

tip

• Unlike bread, panettone remains quite soft when cooked in this way, and care must be taken not to let it break when you turn it in the pan. It is worth the trouble for the wonderful flavor—like really good bread-and-butter pudding, but made in minutes rather than hours.

apricots

Apricots originated in China, where they remain popular. Apart from eating them fresh, Chinese cooks preserve apricots and make a delicious drink from the clarified juice. The fruit traveled to Italy via Armenia and Persia, and was introduced into Britain in medieval times. It did reasonably well in the warmer, southern parts of the country, especially when trained against a wall, but it was in the Mediterranean and North Africa that the apricot really came into its own. It is still an important fruit in Morocco and Algeria, and is often used in savory dishes, such as the tagines that are typical of the area.

In Spain, too, the fruit proved popular, particularly the musk apricot that has a wonderful aroma and very juicy flesh. Spanish missionaries brought it to California, and several varieties are now cultivated in the United States.

Apricots are a good source of beta-carotene, and also contain some vitamin C. They are rich in potassium, and they supply useful amounts of dietary fiber.

plumcots, pluots, and apriums

These are all hybrids, crosses between plums and apricots. What they are called depends upon what percentage of each type of fruit is present. Plumcots are half and half, pluots are more plum than apricot, and in apriums the balance tilts the other way. Some pluots, particularly, are absolutely delicious, with the best attributes of both fruits. Sweeter than most plums, and with better texture, they have an underlying plum flavor with honeyed apricot overtones.

selection and storage

The best apricots are those that have been allowed to ripen on the tree. Look for fruits that are plump, evenly golden in color, and fragrant. They should just yield to the touch, but handle them with care, as they bruise easily. Size is not relevant; some apricots are quite small, resembling ping pong balls, whereas others are twice that size. Apricots on the firm side can be used for cooking, but don't buy really hard specimens that have no aroma. Ripe apricots should be used as soon as possible, but can be kept in a perforated plastic bag in the refrigerator for a few days. Apricots that are very soft should be cooked right away. If you don't want to serve them at once, store the sweetened purée in a plastic tub in the refrigerator, or freeze it.

preparation

There is seldom any need to peel apricots, but if you wish to remove the skins, dunk them in boiling water for a few seconds, then into cold water, as when skinning peaches. Apricots can be cooked whole or halved. Using a sharp knife, cut around the natural seam until you encounter the pit, then twist the halves in opposite directions so that you can separate them. Free the pit with the tip of a sharp knife.

cooking

When apricots are heated, the flavor intensifies. They are delicious cooked, in pies or cakes, and they retain their shape. Apricots have a special affinity for almonds, as the recipe for Apricot Frangipane Cake demonstrates. They also go very well with chocolate, so serving a chocolate sauce with an apricot soufflé or bavarois makes for the perfect partnership. Apricot glaze, made by heating apricot jam with a little water, then pressing the mixture through a fine sieve, is traditionally used on chocolate cakes.

The simplest way to cook apricots is to poach them in a spicy syrup made by heating 1 cup water with ¾ cup sugar and a cinnamon stick. Stir the mixture until the sugar dissolves, then boil the syrup for 2 minutes. Add pitted and halved fruit and poach gently for 8 minutes or until tender. Serve warm or chilled. Apricots are also excellent in savory mixtures. Add a chopped apricot to sausage meat, or use to flavor stuffing for poultry.

drying apricots

Apricots are excellent dried. Skin fresh apricots, halve them, and remove the pits. Place the halves cut side down on baking sheets and dry out in a low oven at 225°F for 4–6 hours. Leave until completely cold, then pack in airtight containers. To use dried apricots, soak them overnight in twice the amount of water by volume, then simmer in the soaking water until tender.

apricot bavarois

There's no better way of using up a glut of apricots than to make this marvelous dessert. Serve it with a dark chocolate sauce for maximum impact.

Serves 6
This recipe contains raw egg, see note on page 142

1 pound apricots, halved
2 tablespoons water
1 tablespoon lemon juice
⅔ cup sugar
½ cup milk
2 tablespoons water
1 tablespoon powdered gelatin
3 eggs, separated
¾ cup heavy cream
whipped cream and chocolate curls, to decorate

Put the apricots in a pan and add the water. Cover and cook gently for 15–20 minutes, until the fruit is soft and pulpy. Remove from the heat and scoop out the pits. Purée the fruit with the lemon juice in a food processor, rub it through a sieve into a bowl, then measure it. There should be 1 cup. If not, add a little fresh orange juice. Stir in 2 tablespoons of the sugar.

Put the rest of the sugar in a heatproof bowl and stir in the milk, with the sweetened apricot purée. Put the water in a cup and sprinkle the gelatin on top. Leave until spongy. Place the heatproof bowl over a pan of simmering water and stir until the sugar has dissolved, then add the gelatin and continue stirring until that has dissolved, too.

In a small bowl, beat the egg yolks with a little of the apricot mixture, then add the mixture to the heatproof bowl. Continue to stir over the heat for about 5 minutes, until the mixture thickens slightly. Leave it to cool and to begin to thicken.

Whisk the egg whites until soft peaks form. Whip the cream lightly. Fold first the cream and then the egg whites into the apricot mixture, and spoon into a 6-cup dessert bowl. Chill for 3–4 hours, until set. Decorate with whipped cream and chocolate curls.

apricot frangipane cake

This cake has a wonderful consistency and a lovely flavor. Combining apricots and almonds is nothing new, but the partnership works so well that it is always worth repeating.

Serves 6

1 stick butter or margarine, softened
½ cup sugar
½ teaspoon almond extract
¾ cup all-purpose flour
1 teaspoon baking powder
½ cup ground almonds
2 eggs
8 ripe apricots, halved and pitted
2 tablespoons apricot jam
2 teaspoons water
¼ cup sliced almonds

Preheat the oven to 350°F. Line a 9-inch round pan with parchment or wax paper. Spray with nonstick cooking spray.

Beat the butter or margarine with the sugar until pale and fluffy. Beat in the almond extract.

Mix the flour, baking powder, and ground almonds in a bowl. In a separate bowl, whisk the eggs until they are pale and thick.

Fold the dry ingredients into the butter mixture, alternately with the whisked eggs. Spoon the mixture into the prepared pan and arrange the apricots, rounded-side up, on top. Bake for 35 minutes.

Cool the cake in the pan for 10 minutes. Carefully remove the cake from the pan, lift off the parchment or wax paper, then place, with the apricot side uppermost, on a wire rack. Leave to cool for 10 minutes more.

Melt the apricot jam with the water in a small pan. Press the mixture through a sieve into a bowl. Brush the top of the cake with the apricot glaze and scatter the almonds on top. Serve in slices.

apricots

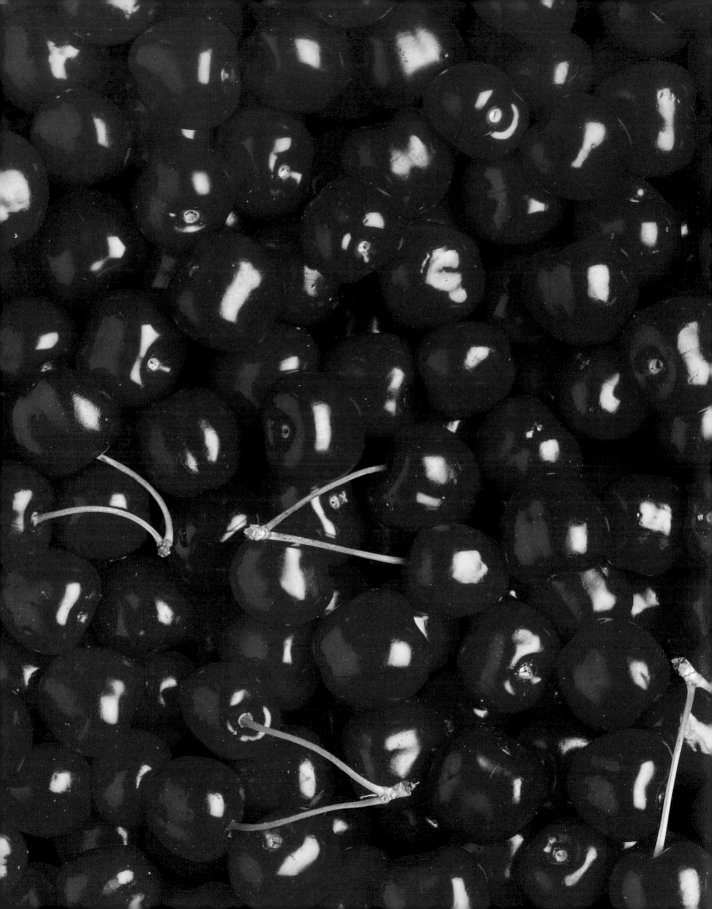

cherries

Although modern hybrids have extended the season, cherries are still only available for a short time during the summer, which is wonderful, for we can really look forward to their arrival each year. These tender, tiny members of the plum family are delicious, whether eaten raw or transformed into soups, sauces, pies, and tarts.

Two types of wild cherry were the forerunners of the hundreds of species available today. The first of these—the sweet bird cherry—grew in Persia and Armenia centuries before the birth of Christ, and was cultivated both by the Chinese and the Medes. Eventually it spread to Europe, where it was received with enthusiasm. At the same time, a bitter wild cherry, the *prunus cerasus*, was making its appearance. Little is known about its origins other than it was first recorded in Greece and Italy, and may have been introduced to Britain by the Romans. Sour cherries, often called pie or tart cherries, are not eaten fresh, but make excellent pie fillings and jams.

French colonists are credited with introducing both sorts of cherry to America. The first commercial cherry orchard was planted in Michigan in 1852, and by the early 1900s the cherry industry was well established. America is now the leading producer of sweet and sour cherries, as well as of hybrids.

sweet cherries

There are basically two types of sweet cherries, those that have crisp flesh and colorless juice, and those with soft flesh and colored juice. Napoleons, which are large and have pale yellow/red fruit, fall into the first category, while the heart-shaped Bing cherries belong to the second. Although they can be cooked, sweet cherries are best eaten fresh.

sour cherries

The best known of these is the Morello, which is round, black, and juicy. Montmorency, a juicy red cherry, is popular in America, as is the Early Richmond.

nutrition

Cherries are a good source of potassium and have antioxidant, anti-inflammatory, and diuretic properties. They contain vitamin A, and are high in dietary fiber.

selection and storage

Buy cherries loose, so that you can inspect each fruit as you pop it into the bag. Choose plump, glossy fruit with bright green stems that are firmly attached. If a lot of the cherries have lost their stems, don't buy the fruit, as it will deteriorate rapidly. Equally, if you spy moldy cherries in a display, buy from elsewhere. Moldy fruits quickly taint their neighbors. Cherries don't keep well, so it is best to buy small quantities that you can eat quickly. This isn't usually a problem, as the fruit is so tasty that you are likely to devour the lot on the way home. If you do need to store cherries for a few days, pack them in a plastic tub, preferably between layers of kitchen paper towels, and keep them in the refrigerator.

preparation

Sweet cherries need no special preparation other than washing. Most people just nibble them off the stem and dispose of the pits discreetly. For cooking, however, you will probably want to pit the fruit, and this is where a cherry pitter comes in handy. This simple device looks a bit like a garlic press. The fruit is placed in a cup and a spiked plunger is driven neatly through it, removing the pit without spoiling the flesh or removing any juice.

cooking

When cooking cherries, try to obtain Morellos or Montmorencies. If these are not available, you can use sweet cherries instead, but the flavor will not be as intense, and you will need to adjust the amount of sugar used. Cherries can be cooked in both sweet and savory dishes. They have a special affinity for duck and game birds. Use cherries in pastries, cakes, and desserts, such as the famous French batter pudding, Cherry Clafouti. Another well-known cherry dish is Cherries Jubilee. The cherries are simply simmered in syrup until tender, then lifted out with a slotted spoon and placed in individual heatproof bowls. The syrup is thickened with arrowroot or cornstarch, then spooned over the fruit. Finally, Kirsch (cherry liqueur) is heated in a metal ladle, then set alight and poured over each dessert.

cherries

cherry soup

Use ripe, flavorsome fruits for this soup—Morellos or Montmorencies are particularly good, but you can use sweet cherries. Serve the soup at the start of the meal or as a dessert, adjusting the amount of sugar accordingly.

Serves 4

3 cups cherries, pitted
1½ cups water
½ cup red wine
½ cup port
1 cinnamon stick
1 long strip of pared lemon zest
2 tablespoons sugar, or to taste
1 tablespoon cornstarch
2 tablespoons water
sour cream or crème fraiche, and toasted, sliced
    almonds, to serve

Put the cherries in a pan and add the water, red wine, port, cinnamon stick, and lemon zest. Bring to a simmer, cover, and cook for 30–45 minutes, until the fruit is tender. Discard the cinnamon stick and lemon zest.

Purée the mixture in a blender or food processor, then rub it through a sieve into a clean pan. Bring to a boil and stir in sugar to taste.

Mix the cornstarch and water in a small bowl. Add to the cherry purée, then stir constantly until the soup boils and thickens. Lower the heat and cook for 5 minutes more, then pour the soup into a bowl and leave to cool, stirring frequently.

Chill the soup for several hours before ladling it into chilled bowls. Spoon a little sour cream on top of each portion, and add a scattering of toasted, sliced almonds before serving.

cherries

duck breasts with cherry sauce

Duck is usually served with oranges, but a cherry sauce works equally well, and looks very attractive on the plate.

Serves 6

6 duck breasts, with skin
salt
fresh thyme sprigs, to garnish

Sauce

6 cups cherries, pitted
1 stalk lemon grass, bruised
¼ cup light brown sugar, or to taste
1 cup orange juice
¼ cup lemon juice
⅔ cup water
1 tablespoon cornstarch
port or cherry brandy to taste (optional)

Start by making the sauce. Put the cherries and lemon grass in a pan and add the sugar. Pour in the citrus juices and ½ cup of the water. Bring to a boil, then lower the heat and simmer for 10 minutes, or until the cherries are soft. Remove the lemon grass.

Mix the cornstarch and the remaining water, then stir the paste into the cherry mixture. Bring to a boil and cook until the sauce thickens. Taste and add a little more sugar, if needed. Keep the sauce warm over the lowest possible heat, stirring occasionally.

Trim the duck breasts. Lay each one in turn on a board, skin side down, and ease away the tendon with a sharp knife. Turn the breasts over and, using a sharp knife, score the skin in a diamond pattern. Rub a little salt into the scored skin.

Heat an enameled, cast-iron grill pan. When it is very hot, place the duck breasts in the pan, skin-side down, pressing them down so that they sizzle and the duck fat starts to flow. Cook for 8–10 minutes, reducing the heat to medium after 2 minutes, then turn the breasts over and cook the other side for a similar length of time, until the meat is as you like it. It is traditional to serve duck pink.

Carve the duck breasts at an angle into thin slices. Fan out the meat on heated plates. Taste the cherry sauce and add a splash of port or cherry brandy if you like. Spoon a little of the cherry sauce onto each portion of duck. Garnish with the thyme. Pour the remaining cherry sauce into a bowl and offer it to anyone who wants extra.

tips

• Use a spatula to press the duck flat on the pan while cooking, and pour away the excess fat occasionally.
• The precise timing will vary, depending on the type of duck breasts used, their thickness, and the heat of the burner.

cherries

cherry clafouti

In France, this is a favorite dessert during cherry-picking time. It tastes delicious.

Serves 4

2 tablespoons confectioners' sugar,
    plus extra for dusting
3 cups sweet cherries, pitted
¾ cup self-rising flour
½ cup ground almonds
2 tablespoons sugar
4 tablespoons butter, melted
⅔ cup milk
2 eggs

Preheat the oven to 375°F. Grease a 9-inch shallow baking dish thoroughly and dust the bottom of it with the measured confectioners' sugar. Arrange the cherries on top.

Mix together the flour, ground almonds, and sugar in a bowl. Pour the melted butter into a pitcher and whisk in the milk and eggs.

Make a well in the center of the flour mixture and add the butter mixture. Stir it well, gradually incorporating the flour, to make a smooth batter.

Pour the batter over the cherries and bake for 25 minutes. Run a knife around the rim of the dessert and invert it onto a flat serving platter. The top of the clafouti (formerly the bottom of the dessert) should be creamy and studded with cherries, while the rest of the mixture should have set to a sponge. Dust the top with confectioners' sugar and serve.

tips
• You can leave the pits in the cherries, if you like, but they can be a bit of a nuisance.
• If you do not have self-rising flour, use all-purpose flour plus 1 teaspoon baking powder.

cherries

figs

The sweetest of all fruits, figs are truly luscious, especially when eaten straight from the tree. They are believed to have originated in Syria, and are mentioned in Babylonian texts that date back 4,000 years. In the Garden of Eden, fig foliage was used to preserve modesty, a fact that was commemorated when fig growers needed a name for a new variety with particularly large leaves. They called it Adam.

Figs were venerated by the ancient Greeks. It was illegal to take them out of the country, and the word sycophant ("one who shows the fig") was coined to describe those who revealed the whereabouts of the fruit to fig smugglers. The Romans dedicated the fig to Bacchus, and offered the god the first fruits of every season.

Throughout the Mediterranean, figs grow in profusion, and are served fresh and dried. In Central Europe roasted figs are used to flavor coffee. Italians serve fresh figs with prosciutto, and the French douse them in Cointreau.

Figs were first planted in Britain during the reign of Henry VIII, and they continue to be grown in southern parts of the country, despite the fact that the climate is not really suitable. They do better in America, particularly in California, where they have been cultivated since the Spanish missionaries introduced them in the 18th century. The Mission fig—a honey-sweet, deep-purple variety—celebrates this connection.

varieties

Figs are often described by color. There are white, red, and black (purple) varieties, all with red flesh edged with white. American favorites include the Calimyrna—a succulent white fruit—and the Kadota, a creamy yellow or green fig which is also popular in Greece and Italy. Two of the figs that grow in Britain are the brownish-red Brown Turkey and the pale green White Marseilles.

nutrition

Figs are a good source of vitamin A. They are high in soluble fiber, and yield potassium and calcium. Their laxative properties are well documented.

selection and storage

If possible, buy figs that have been allowed to ripen naturally on the tree. Look for fruits with whole, unblemished skins. They should be tender and yielding, but retain a hint of resistance. If they smell sour or feel mushy, they are overripe and will taste unpleasant. Handle figs with care, as they can easily become squashed or bruised. Eat them as soon as possible. If you must store them, they will keep for a few days in the refrigerator, preferably in a plastic container lined with paper towels. Do not put the lid on the box, simply cover the figs with more paper towels. Let the figs return to room temperature before serving them, as chilling them diminishes their flavor.

preparation

Washed figs do not need any preparation other than the removal of the stem. The entire fruit is edible. They look very pretty if you cut them into flower shapes. This is easy to do: simply stand a fig upright on a board and use a sharp knife to cut it into quarters, stopping short just before you get to the base. Press up gently from underneath and the quarters of fig will open out like petals. You can drizzle a little orange juice, liqueur, or brandy over the fruit and macerate it before serving. Alternatively, fill the center with mascarpone or whipped cream.

cooking

Whole or sliced figs can be baked, but taste best when they are simply warmed through. They are delicious when lightly grilled with goat's cheese, feta, or haloumi, and they also have an affinity for nuts and chocolate. Chopped figs are a good natural sweetener, and can be used in cakes and bakes. Fig ice cream has an unusual, almost nutty flavor, and is becoming increasingly popular.

fig & goat cheese bruschetta

Serve this simple dish for a light lunch or as a sensational starter. It looks gorgeous, and the combination of melted goat cheese and warm figs with just a hint of mango chutney is irresistible.

Serves 6

6 ripe purple figs
12 ounces full-fat goat cheese
1 small ciabatta or similar loaf, about
    10 inches long
¼ cup mango chutney
salt and pepper
salad leaves, to garnish

Slice the figs thinly lengthwise. Cut off and discard the rind from the goat cheese, then slice it.

Preheat the grill. Cut the loaf of bread in half lengthwise and spread both halves with mango chutney. Arrange alternate slices of fig and goat cheese down the length of both pieces of bread, making two parallel rows on each, and starting the second row with goat cheese. Season with salt and plenty of black pepper.

Grill for about 3-4 minutes, until the cheese has melted a little and begun to brown around the edges. Slice each piece of topped ciabatta widthwise in three, and place a piece on each serving plate. Garnish with salad leaves and serve immediately.

fig & armagnac ice cream

The riper and more flavorful the figs, the better this sophisticated ice cream will taste.

Serves 4–6

1 pound ripe fresh figs
3 tablespoons Armagnac or sherry
4 egg yolks
¾ cup sugar
1¼ cups milk
1¼ cups whipping cream
1 teaspoon pure vanilla extract

Break off the stem end of each fig, then cut them into quarters. Put them in a food processor with the Armagnac or sherry, and process to a purée.

Beat the egg yolks with ½ cup of the sugar in a bowl. Heat the milk until almost boiling, then whisk it into the beaten egg yolks. Pour the mixture into the top of a double boiler, set it over simmering water, and cook, stirring constantly until the custard thickens enough to coat the back of the spoon. Leave to cool, stirring occasionally.

When the custard is cold, stir in the puréed figs and the cream, with the vanilla extract. Taste the mixture and add the remaining sugar if needed.

Chill the mixture, then churn in an ice cream maker, or freeze in a shallow container, whisking in the ice crystals at regular intervals. Let the ice cream soften a little before serving. Serve in scoops, with fresh fig slices.

tips
• Choose a round, full-fat goat cheese that is firm enough to slice. Fresh mozzarella could be used instead, but will lack the delicious sharpness that makes the goat cheese such a fine foil for the figs.
• To make the cheese easier to slice, chill it briefly in the freezer.

persimmons

Persimmons can be delicious—but only if they are absolutely ripe, or belong to one of the new varieties from which the natural tannin has been eradicated. If you encounter a type of persimmon that is not tannin-free, and attempt to eat it before it has reached its prime, it will probably taste foul. Wait a while, and the same fruit could easily taste wonderful.

Persimmons have been cultivated for centuries in China and Japan, and are now produced commercially in Israel, Italy, France, and Spain. There are numerous varieties, including one that grows wild in America. The wild fruit is very small, which accounts for its common name, the Virginian date, and is so astringent that it is virtually inedible until after the first frosts. Americans tend to prefer the Japanese persimmon—the Hachiya—which is big and round, with a pointed end. Hachiyas do contain tannin, but when allowed to ripen and soften, this becomes less conspicuous and the creamy flesh develops a delightful sweet-sour flavor. Even sweeter is the Fuyu, a tannin-free persimmon,

that can be eaten when still quite firm. Fuyus are smaller than Hachiyas, and are shaped rather like tomatoes.

The Sharon fruit is a very well-known variety of persimmon that was developed by the Israelis. Not only is it tannin-free, but it also lacks the inedible brown seeds you will find in some of the other varieties. Sharon fruit has very thin skin and firm, jelly-like flesh which tastes delicious. A little lime juice brings out the flavor.

nutrition
All persimmons contain large amounts of vitamin A, and are a source of potassium, calcium, and iron.

selection and storage
Look out for orange-red fruits that are plump and glossy. The skin should be smooth, unbroken, and bright, almost translucent, and the caps (calyxes) and stems should be attached. Tote a fruit in your hand; it should feel heavy for its size. Fuyus and Sharon fruit can be eaten when they are on the firm side, but all other types should be very soft. Ripe fruit can be stored in the refrigerator for up to 3 days. Unripe fruit will probably ripen eventually if placed in a brown paper bag with a banana, but may have a disappointing flavor.

preparation

As long as the skin is sweet and tender, as it is likely to be on a Sharon fruit or Fuyu, it can be eaten with the flesh. Just cut away the cap. If the skin tastes bitter, or is thick, slice the fruit in half from top to bottom, then scoop out the flesh with a teaspoon.

cooking

Persimmons can be used in cakes and puddings, but are best eaten fresh. Puréed persimmon flesh, spiked with lime or lemon juice, makes a good sauce. It is also excellent in ice cream, especially if you mix it with passion fruit.

persimmon & passion fruit ice cream

This has a beautiful pale primrose color and tastes lovely, especially if you allow the ice cream to thaw slightly before serving it. This process, called "ripening," lets the full flavor of the ice cream develop.

Serves 4

3 ripe persimmons
3 passion fruit
juice of 1 lemon
5 tablespoons sugar
1¼ cups whipping cream
extra passion fruit pulp to decorate (optional)

Cut off the top of each persimmon and spoon the flesh into a bowl, then use a teaspoon to scrape as much flesh off the skins as possible. Add it to the bowl. Place a sieve over the bowl. Cut each passion fruit in half. Scrape the pulp into the sieve, then press it through with the back of a spoon, leaving the black seeds behind.

Spoon the mixture into a food processor, add the lemon juice, and process to a fine purée. With the motor running, gradually add the cream until well combined. Scrape the mixture into a bowl, chill it, then churn it in an ice cream maker. Alternatively, freeze in a shallow container, whisking in the ice crystals at regular intervals. Before serving, let the ice cream soften slightly. Serve in scoops, with a little passion fruit pulp spooned over.

pomegranates

The name means "apple with many grains," a reference to the numerous seeds, each encapsulated in sweet juice, that lie inside the fruit. Pomegranates are not particularly pretty, so it comes as a surprise when you first cut one open and discover the mass of tiny pale pink bubbles that make up the interior. These bubbles, or arils are they are correctly known, are surrounded by inedible white pith that must be removed before the juice and seeds can be enjoyed.

The sheer number of the seeds led to the fruit's being regarded as a fertility symbol in the Middle East, where it originated. Before long it was being cultivated in China, India, and throughout the Mediterranean, and even found its way across the sea to North Africa, where the skin, that contains tannin, was used in the preparation of hides for the leather industry. Pomegranate juice stains, and was used by the Persians to dye wool for their famous carpets. The juice is also used to make drinks.

A pomegranate syrup, grenadine, is a popular ingredient in cocktails and desserts.

The fruit continues to be grown throughout Asia and the Mediterranean region, but is also widely cultivated in California. Its use as a salad ingredient has been hailed as innovative and exciting, but it is actually nothing new—a 16th century menu reprinted in *Larousse Gastronomique* includes a pomegranate salad!

Pomegranates are rich in potassium and contain a fair amount of vitamin C. They are a good source of dietary fiber.

selection and storage

Choose fruits with bright, fresh color. The skin should have no blemishes or cracks. Squeeze the brownish "crown" on top of the fruit; it should give a little under your fingers. Pomegranates keep very well and can be stored in a cool, dark cellar for up to 2 months, or in a perforated bag in the refrigerator for up to 3 months.

preparation

There are several ways of preparing a pomegranate. Whichever one you choose, wear an apron, as the juice stains. One method involves cutting the fruit in half and lifting out each aril in turn with a pin. This can be time-consuming, so a better solution is to slice the top off the pomegranate, just below the "crown," then score the skin from top to bottom in several places, dividing the fruit into eight segments. Working over a bowl to catch the juice, peel back the skin from each segment and scrape the arils into the bowl, trying not to include any of the bitter white pith or skin. Some cooks put the segmented fruit in a bowl of water before breaking it up. That way, the arils sink and everything else floats and can easily be skimmed off. The arils can then be drained and dried.

pomegranate juice

If all you want is the pomegranate juice, either cut the fruit in half and squeeze it on a lemon squeezer, or scrape out the arils in a metal sieve and use a spoon to burst them and release the juice. Be gentle, though; if you crush the seeds too much, the juice will taste bitter.

cooking

Pomegranates are best eaten fresh. The arils are delicious just as they are, or with a little liqueur poured over them. They look and taste wonderful in green salads.

pomegranate syrup

This dark syrup is made from concentrated pomegranate juice (and is sold as such in the U.S.). It has a wonderful, rich flavor and can be used in dressings, as an alternative to the ubiquitous balsamic vinegar. It makes a great glaze on grilled meat or fish, and a dash, added to stews or casseroles, deepens the flavor. The dilute syrup is used in drinks and to make sorbets.

salad leaves with pomegranate dressing

This must be one of the prettiest salad dressings around. It has a delicate flavor that is easily overwhelmed, so do not be tempted to use virgin olive oil or balsamic vinegar.

Serves 6

sufficient mixed salad leaves for six
Dressing
1 pomegranate
1½ teaspoons red wine vinegar
1 tablespoon light olive oil
salt and pepper

Make the dressing. Cut the pomegranate in half. Using a citrus squeezer, squeeze one half of the pomegranate and tip the juice into a small bowl. Add the vinegar and oil, with plenty of seasoning, and whisk well.

Spoon the arils from the remaining pomegranate half into a bowl, taking care not to include any of the surrounding white pith. Stir the arils into the dressing, and chill.

When ready to serve, tip the mixed leaves into a large salad bowl. Add a little of the dressing and toss well so the leaves are well coated. Then spoon more dressing over the salad so that the pink of the pomegranate can clearly be seen against the dark green leaves. Serve the remaining dressing separately.

pomegranate and shrimp risotto

Fruit risottos are becoming increasingly fashionable, and this combination works extremely well. The flavor is sensational with the pomegranate arils providing a crisp contrast to the creaminess of the rice.

Serves 4

2 pomegranates
4 cups chicken stock
3 tablespoons olive oil
1 onion, finely chopped
1 celery stalk, sliced
2 cups arborio or other risotto rice
½ cup white wine
14 ounces peeled, cooked, large shrimp,
    thawed if frozen
salt and pepper
1 tablespoon butter
shavings of Parmesan cheese (optional)
lime segments, to serve

Cut one pomegranate in half. Using a citrus squeezer, as when squeezing an orange, squeeze both halves and tip the juice into a small pitcher. Cut the remaining pomegranate into quarters and spoon the arils into a bowl, taking care not to include any of the surrounding yellow pith.

Heat the chicken stock in a pan. Let it simmer gently, ready for adding to the risotto. Heat the oil in a large, wide pan, and fry the onion and celery until soft, but not colored. Stir in the rice until all the grains are coated in oil.

When the mixture is hot, add the wine and stir until it has been absorbed by the rice. Start adding the hot stock, a ladleful at a time. Stir constantly until each addition has been completely absorbed.

When about a quarter of the stock has been incorporated, pour in the pomegranate juice, and continue to stir until all of it has been absorbed.

Continue to add the stock, never more than a ladleful at a time, and patiently stir all the while. This is the secret of a good risotto.

When you have been stirring for about 18 minutes, most of the liquid will have been absorbed and the rice will be creamy, but still fairly firm in the center of the grain. Add the shrimp and continue to stir over the heat, adding the remaining stock if needed, until the shrimp have heated through and the center of each grain of rice provides just a hint of "bite." At this point, add salt and pepper to taste.

Stir in the butter, with a little Parmesan if you like, and cover the pan. Leave the risotto to stand for 1 minute, then stir in the pomegranate arils. Serve at once, offering your guests lime segments so that they can squeeze a little juice over the risotto if they like. You could also offer more Parmesan at the table, but the risotto is so delicate that it really doesn't need it.

tip
• You may not need to add all the stock. The best way to judge if the rice is perfect is to taste a grain. When it is creamy and tender, but still *al dente* in the center, it is ready.

berries

You could call it "berried" treasure, that glorious collection of jewelled fruits that see us through summer and early autumn. Soft and yielding strawberries, tangy and tempting raspberries, gooseberries, cranberries, blueberries, blackberries, and currants—the list goes on and on.

Berries are essentially wild fruit, and although many varieties are now cultivated, there is still huge pleasure to be had from hunting in the hedgerows for blackberries or bilberries, and bearing home baskets laden with delicious fruit that you haven't had to pay for.

One of the earliest signs of summer is when the pick-your-own fruit farms start advertising their wares. Sun-warmed strawberries, grown in raised beds so you don't even have to bend down to pick them; loganberries hiding under glossy leaves in long, cool corridors of green; plump gooseberries, dusky blueberries, clusters of black, red, and white currants: they are all there for the picking. All berries contain vitamin C. Cranberries and black currants are particularly good sources.

strawberries

Perhaps the best loved of all the berries, these began as small, wild woodland fruit. Fraises du bois, and the slightly larger Alpine strawberries, can still be found in the wild, but are also cultivated. Both have excellent flavor and perfume. It was an American wild strawberry, crossed with a variety from Chile, that gave rise to the familiar heart-shaped strawberry that is so widely cultivated today.

When picking or buying strawberries, look for fruit that is ripe but still firm enough to handle. They should be brightly—and evenly—colored, and there should be a discernible strawberry aroma. Smaller berries are often more intensely flavored than large ones.

Strawberries are so good that the best way to eat them is on their own, preferably straight off the plant, but they can also be used in fruit or vegetable salads, on shortcakes, or in baked pastry cases, with cream. Conventional cooking robs them of their color and texture, but they make very good jam.

raspberries

Summer is well established when the first raspberries appear, their refreshing flavor a rich reward for the pickers. Sometimes called hindberries, these delicate fruit are sweet, yet tangy. Purists love them just as they are, with neither sugar nor cream, but they can be baked in pies or tarts, or used as luscious fillings for cakes or crepes. Raspberry Mousse is light and lovely, and raspberry coulis, made by rubbing puréed raspberries through a sieve, then stirring in confectioners' sugar, is a delicious sauce. Because they are tart, raspberries are also good with any food that would normally be enhanced by lemon or lime. A raspberry sauce goes very well with roasted fish.

Most raspberries are a beautiful clear red, but you can occasionally find white or yellow varieties. The salmonberry is a wild American raspberry. As the name suggests, it is usually salmon pink, but can also be burgundy in color. Arctic cloudberries are golden members of the raspberry family. They grow in the far north of Scandinavia. The sweetly named honeyberry is another Arctic berry. Most of the crop, which is very small, goes to making a delicious liqueur called Mesimarja.

blackberries

These dark, glossy fruit grow wild on brambles and picking them can be a prickly problem, so it is perhaps fortunate that growers have now developed thornless varieties. If they have had enough sun, the wild fruit can be sweet enough to eat raw, but they are more often baked in pies or puddings. Blackberries are rich in pectin and make wonderful jelly and jam. The fruit resembles the raspberry, but is solid through the center, whereas raspberries are hollow once picked, as the core is generally left behind on the cane.

Blackberries have been cultivated in America for more than a century. The marionberry is a particularly fine variety. Look out for blackberry and raspberry hybrids such as youngberries, tayberries, loganberries, and boysenberries too. Trailing blackberries are called dewberries.

gooseberries

Large and luscious, these green, red, or golden globes grow on low bushes and are native to northern Europe. They are nowhere more popular than in Britain, where they are used in both savory and sweet dishes. Grilled Mackerel with Gooseberry Sauce is a classic dish, as is gooseberry fool, a simple dessert consisting of puréed gooseberries stirred into a mixture of cold custard and whipped cream. Some varieties of gooseberry are sweet enough to eat raw, but most are very tart, and must be cooked with plenty of sugar.

cranberries

Brimming with vitamin C, cranberries are delicious cooked in compotes or sauces, but are usually too bitter to eat raw. A small variety of cranberry grows wild in Finland, which is also home to the tiny lingonberry. Most of the world's cranberries come from the United States, however, where they have been growing wild in the marshy areas of New England for centuries. They were originally known as craneberries.

cranberry sauce

Put 2 cups cranberries in a pan and add 1 cup sugar. Using a zester, pare off strips of orange rind and add them to the pan, then squeeze the orange and add ½ cup of the juice to the cranberry mixture. Simmer until the sugar has dissolved and the berries have popped. Serve warm or cold.

berries

54

Cranberries are high in vitamin C and also yield vitamin A and potassium. They contain plenty of pectin, so jams and jellies made from the berries set well. The fruit tastes good in both sweet and savory dishes. Cranberry juice is a popular and nutritious beverage, which is claimed to be effective in treating urinary tract infections.

blueberries

These slate-blue berries have a mild, sweet flavor and can be eaten raw. They look very attractive on the plate, and are often used as a garnish, with a light dusting of confectioners' sugar. Fanned slices of mango look particularly pretty when served this way. Blueberries also make wonderful, juicy pies. Try them in a one-crust pie with a streusel topping to which you have added some almonds—a perfect partnership. Wild members of the blueberry family include bilberries, huckleberries, whortleberries, blaeberries, and tangleberries. When buying or picking blueberries, look for firm, plump fruit with tight skins. A slight dusty quality to the skin is normal, and proves that the berries are fresh. Blueberries freeze well, and can be used straight from frozen in some recipes. They make marvelous muffins and pies.

currants

Miniature members of the gooseberry clan, these can be red, white, or black. They grow in clusters.

Black currants are rich in vitamin C; red or white currants somewhat less so. Look for bright, shiny fruits with no sign of mold. The berries are often sold on the stalks. To remove them, run the tines of a fork between the fruit and the stalk and they will drop off. If you intend to freeze the currants, you can leave the stalks on. When they are solid, shake the frozen stalks over a bowl and the berries will obligingly fall in. Currants are delicious in pies and tarts, especially if you use the more distinctively flavored black fruit. Red and white currants are very pretty, and are often used to decorate desserts. All currants are rich in pectin, and make excellent jams and jellies.

selecting and storing berries

When you pick your own fruit, handle it with care. Berries are delicate and easily crushed. The same applies to bought fruit. Buy small quantities in clear plastic boxes if possible, so you can see at a glance that the ones on the bottom are unbroken. Check them more thoroughly when you get home and discard any that are very soft. Keep them in the refrigerator, in the original container, for no longer than three days. Do not wash or hull them (in the case of strawberries) until you are ready to serve. Moisture hastens decay, so make sure the fruit is dry. When ready to serve, wash them briefly. Let strawberries come to room temperature before serving.

strawberry & watercress salad

This salad celebrates the classic combination of strawberries and balsamic vinegar, setting them against a background of dark green, peppery watercress leaves and crisp celery. Serve the salad solo or try it with smoked shrimp or ham.

Serves 4

1 bunch watercress
2 celery stalks, sliced
1 small mild red onion, sliced in rings
1 pint just-ripe strawberries

Dressing

2 tablespoons balsamic vinegar
pinch of sugar
1 tablespoon very finely chopped shallot
6 tablespoons extra virgin olive oil
salt
½ teaspoon drained pink peppercorns in vinegar

Trim off any tough stalks from the watercress, wash it well, then drain it and pat the leaves dry with paper towels.

Mix the watercress, celery, and onion slices in a bowl. Hull the strawberries, slice them lengthwise and add three-quarters of the slices to the bowl.

Make the dressing. Put the balsamic vinegar in a small bowl, and add the sugar and shallot. Whisk to mix, then gradually whisk in the olive oil. Add salt to taste, then stir in the pink peppercorns.

Pour 2 tablespoons of the dressing over the salad and toss lightly to coat the watercress and strawberries. Spoon onto salad plates, mounding the salad in the center. Arrange the remaining strawberries on top, and serve. Offer the remaining dressing separately.

tip

• Some people are allergic to pink peppercorns, so always inform your guests that you have used them.

strawberry shortcake

An American classic, this is a very good way to serve what many people claim is their favorite fruit. Decorate the plate with strawberry leaves, if you have any.

Serves 6–8

2½ cups all-purpose flour
1 tablespoon baking powder
½ teaspoon salt
1¼ sticks butter
¼ cup sugar
2 eggs, beaten

Filling

3 pints strawberries
½ cup sugar
⅔ cup heavy cream

Start by making the filling. Set two of the best strawberries aside for the decoration. Hull the remaining strawberries. Slice half of them and roughly chop the rest. Put them in separate bowls and divide 5 tablespoons of the sugar between them.

Make the shortcake. Preheat the oven to 400°F. Grease a 9-inch round cake pan. Mix the flour, baking powder, and salt in a bowl. Cut in the butter, then rub it in until the mixture looks like fine breadcrumbs. Stir in the sugar. Add the eggs and mix to a soft dough. Pat it out to a round, ease it into the prepared cake pan, and flatten the surface.

Bake for 30 minutes. Turn out the shortcake on a wire rack and let it cool for 3–4 minutes. Meanwhile, whip the cream for the filling with the remaining sugar.

Cut the shortcake in half to make two layers. Place the bottom layer on a serving dish, taking care not to break it. Spoon over the chopped strawberries, along with any juice, then top with half the sliced strawberries. Replace the top layer and leave the shortcake to cool slightly.

Decorate the cake with the whipped cream and the remaining sliced strawberries, arranging them around the rim of the shortcake. Fan the reserved whole berries by slicing them lengthwise from the tip, stopping the knife short of the base so that they are held together. Arrange these strawberries in the center of the top of the shortcake, spreading out the fruit decoratively. Serve in slices while still warm.

tip

• It is traditional to spread the shortcake with butter before adding the strawberry filling. In this recipe, however, the butter has been omitted to reduce the level of fat, and the results are still very good.

strawberries romanoff

When strawberries are at their peak, their intense flavor is such that the best way to eat them is just as they are. Berries which have been in cold storage sometimes need a helping hand, and this is a very good way to serve them.

Serves 4–6

6 sugar cubes
1 orange
6 tablespoons Curacao or Grand Marnier
2 pints strawberries, hulled
Crème Chantilly
1¼ cups heavy cream
1 tablespoon confectioners' sugar
½ teaspoon natural vanilla extract

Rub the sugar cubes over the surface of the orange to impregnate them with the oil. Put the cubes in a bowl. Squeeze the orange, and strain the juice into the bowl. Crush the sugar cubes in the juice, then stir in the Curacao or Grand Marnier. Add the strawberries, stir gently, then cover and chill for 2 hours. Stir occasionally, but take care not to break up the fruit.

Make the Crème Chantilly by whipping the cream, then stirring in the confectioners' sugar and vanilla. Put it in a piping bag fitted with a star nozzle.

Drain the strawberries. Pile them up, pyramid fashion, on a serving dish, preferably one which is on a stand, and decorate by piping the cream in the gaps and around the pyramid.

variation
• For sheer indulgence, make small meringues, roughly the same size as the strawberries, and add them to the pyramid with the cream.

strawberry jam

Strawberries do not contain much pectin, so you need a little lemon juice to ensure a good set.

Makes about 9 8-ounce jars

3 pounds just-ripe strawberries, hulled
¼ cup lemon juice
6 cups sugar

Put about three-quarters of the strawberries in a heavy saucepan. Working over the pan, nick each of the remaining strawberries with a sharp knife, to encourage the juice to flow, and add them to the pan. Pour in the lemon juice. Heat gently for 10 minutes or until the strawberries are soft, stirring frequently, then sprinkle over the sugar and stir gently until it has dissolved.

Bring the mixture to a boil, and boil rapidly until the setting point is reached. If you are using a sugar thermometer, it should read 220°F. If not, test by spooning a little of the jam onto a saucer which has been chilled in the refrigerator. Leave for a minute or two to cool, then gently push the jam with your fingertip. If the surface wrinkles, the jam is ready.

Switch off the heat. Skim off any scum from the surface of the jam, then leave it to stand for 15 minutes. Stir and ladle into hot jars. Seal with canning lids and process in a boiling-water bath for 5 minutes.

tip
• If you warm the sugar in a bowl in the oven before it is added to the strawberries, it will dissolve much more readily, and will not lower the temperature of the mixture. The jam will consequently reach setting point sooner.

berries

roast cod with raspberry sauce

It is odd how happily we accept lemon or lime with fish, but seldom consider using other types of fruit. This sweet and sour raspberry sauce tastes wonderful with cod.

Serves 4

1½ pints raspberries
¼ cup water
2 tablespoons light brown sugar
1 tablespoon balsamic vinegar
salt and pepper
1 teaspoon black peppercorns
½ stick butter, softened
4 6-ounce portions cod fillet, skin on
2 tablespoons sunflower oil
fresh mint leaves, to garnish

Preheat the oven to 375°F. Set aside a quarter of the raspberries for the garnish and put the remainder in a small pan. Add the water, sugar, and balsamic vinegar. Bring to the simmering point, then cover the pan and cook the raspberries for 5 minutes, shaking the pan occasionally.

Rub the mixture through a sieve into a bowl. Season with plenty of salt and pepper, then return to the pan and heat gently.

Crush the peppercorns roughly. Beat the butter in a small bowl until soft, then add the peppercorns. Spread the mixture over the flesh on all four pieces of cod.

Heat the oil in a frying pan that can be used in the oven. When the oil is hot, add the cod pieces, skin side down. Cook for 2–3 minutes, spooning the hot oil over the top of the fish from time to time.

Transfer the pan to the oven and cook for 6–8 minutes more or until the fish flakes when tested with the tip of a sharp knife.

Spoon a little of the raspberry sauce onto each plate, add a piece of the roast cod, and garnish with the reserved raspberries and mint leaves. Serve the remaining raspberry sauce separately.

raspberries

raspberry crepes

If you make the crepes ahead of time, this is a very quick and easy dessert.

Serves 4

Crepe batter

1 cup all-purpose flour

pinch of salt

2 teaspoons sugar

1 egg, beaten

1¼ cups milk

2 tablespoons oil

1 tablespoon butter

2 tablespoons light brown sugar

4 scoops vanilla ice cream, to serve

Filling

1½ pints raspberries, hulled

¼ cup sugar

3 tablespoons orange juice

2 tablespoons Cointreau

Mix together the flour, salt, and sugar in a bowl. Make a well in the center. Add the beaten egg and milk to the well. Stir the liquid, then gradually incorporate the surrounding flour mixture to make a smooth batter. Stir in half the oil, and set aside for 20 minutes.

Meanwhile, prepare the filling. Put the raspberries in a bowl. Sprinkle over the sugar, and pour over the orange juice and half the Cointreau. Set aside for 20 minutes.

Make the crepes. Heat the remaining oil in a 6-inch crepe pan. Tip the excess oil into a small pitcher, leaving just enough to grease the pan. Add a generous spoonful of batter and tilt the pan so that it covers the bottom evenly. Cook the crepe for about 30 seconds.

Once the crepe has set, shake the pan gently to keep it on the move. When it is brown underneath, flip it over and cook the other side briefly. Slide it onto a plate and cook seven more crepes in the same way.

Using a slotted spoon, divide the raspberries among the crepes and roll them up.

Melt the butter in a large frying pan and add the brown sugar. Stir over low heat until dissolved, then stir in the liquid in which the raspberries were soaked.

Carefully add the rolled crepes to the pan and heat through for 2–3 minutes. Pour the remaining Cointreau over the crepes. Serve immediately, with ice cream.

tips

• If the batter is at all lumpy, don't despair. Just strain it through a sieve into a clean bowl.

• Crepes freeze well. Simply stack them, placing pieces of wax paper between the layers, then wrap them in foil and freeze. Bring to room temperature before using.

raspberry mousse

Light and creamy, with a sauce that is just sufficiently tart to provide a refreshing contrast, this is a delicious dessert for late summer.

Serves 6

2 pints fresh raspberries
2 tablespoons confectioners' sugar
2 tablespoons lemon juice
¼ cup water
1 tablespoon powdered gelatin
2 eggs, plus 1 egg yolk
6 tablespoons sugar
1 cup heavy cream
whipped cream, to decorate

Set aside a few raspberries for decoration. Put the rest in a food processor or blender and add the confectioners' sugar and lemon juice. Process to a purée, then remove the seeds by pressing the purée through a sieve placed over a bowl.

Pour the water into a metal measuring cup and sprinkle the gelatin on top. When it is spongy, set the cup over simmering water until the gelatin melts.

Meanwhile, place the eggs, egg yolk, and sugar in the top of a double boiler. Whisk over simmering water until the mixture is very thick, then remove from the heat and gently fold in the gelatin mixture, with 1 cup of the raspberry purée. Save the remaining purée for a sauce.

Whip the heavy cream to soft peaks and fold it into the mixture. Spoon into a glass bowl and chill for 3–4 hours, until set.

Decorate with whipped cream and the reserved raspberries. Taste the raspberry sauce and add a little extra sugar if you like. Serve with the mousse.

blackberry & apple brown betty

Whether you pick your own blackberries, or buy cultivated varieties, this is absolutely delicious, the crisp crumb topping providing the perfect contrast to the juicy fruit.

Serves 6

2 pints blackberries, hulled
1 orange
1 pound cooking apples
½ teaspoon ground cinnamon
1 cup sugar
5 slices whole wheat bread, crusts removed
4 tablespoons butter
cream, ice cream, or crème fraiche, to serve

Preheat the oven to 275°F. Put the blackberries in a baking dish. Using a zester, remove the zest from the orange and put it in a shallow bowl. Cut the orange in half. Squeeze one half and add the juice to the bowl. (Save the remaining orange for another recipe.)

Quarter, core, and peel the apples, then slice them into the bowl of juice. Toss the apple in the juice so they are thoroughly coated (this will prevent them from discoloring), then add the mixture to the blackberries. Sprinkle over the cinnamon and half the sugar, toss well and set aside.

Crumb the bread in a food processor. Melt the butter in a pan, and stir in the breadcrumbs and remaining sugar. Spoon the mixture over the fruit, gently level the surface, and bake for about 35–45 minutes, or until the crust is golden and the apples are tender.

Serve hot, with cream, ice cream, or crème fraiche.

broiled mackerel with gooseberry sauce

This classic combination is particularly popular in France. The sauce goes well with any oily fish, and also tastes wonderful with pork chops or baked ham.

Serves 4

4 10-ounce mackerel, cleaned, fins removed
salt and pepper
½ cup olive oil
¼ cup lemon juice
1 garlic clove, crushed
1 tablespoon chopped fresh oregano
2 tablespoons chopped fresh parsley

Sauce

2½ cups gooseberries, trimmed
1 tablespoon grated fresh ginger
¼ cup water
2 teaspoons sugar
1 tablespoon butter

Make three slits on both sides of each mackerel. Season well, then lay them in a non-metallic dish.

Mix the oil, lemon juice, garlic, and herbs, and pour the mixture over the fish. Marinate for 1–1½ hours, turning occasionally.

Meanwhile, make the sauce. Put the gooseberries, ginger, and water in a pan. Cover and simmer until the gooseberries are soft, then tip the mixture into a sieve set over a clean pan, and rub it through with the back of a wooden spoon.

Preheat the broiler. Lift the mackerel out of the marinade and place them in the broiler pan. Broil for 5–8 minutes on each side, until the flesh flakes when tested with the tip of a sharp knife.

When the fish is almost ready, reheat the gooseberry sauce and beat in the sugar and butter. Serve with the fish.

red gooseberries

white currant & gooseberry jam

White currants are nowhere near as easy to locate as black currants or red currants, but they are worth seeking out, as they look very pretty when frosted as a decoration. They also make marvelous jam.

Makes about 11 8-ounce jars

8 cups sugar
2 pounds white currants,
    stripped from stalks
2 pounds gooseberries, trimmed

Put the sugar in a heatproof bowl and place it in a low oven, 300°F, to warm while you start cooking the fruit.

Put the currants and gooseberries in a preserving pan. It is not necessary to add water. Heat very gently, pressing the gooseberries against the sides of the pan with a wooden spoon until they start to break down and give up their juice.

Sprinkle over the sugar, and stir gently until it has dissolved.

Bring the mixture to a boil and boil rapidly until the setting point is reached. If you have a sugar thermometer, it should read 220°F. If not, test by spooning a little of the jam onto a saucer which has been chilled in the refrigerator. Leave for a minute or two, then gently push the jam with your fingertip. It should wrinkle.

Switch off the heat. Skim off any scum from the surface of the jam, then leave it to stand for 15 minutes. Stir and ladle into hot jars. Seal with canning lids, and process in a boiling-water bath for 5 minutes.

green gooseberries

berries

65

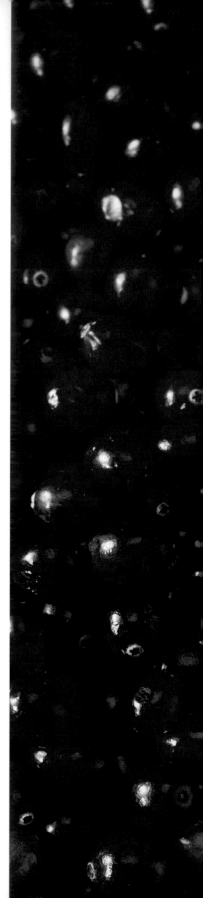

barbecued spareribs with cranberries

Ribs taste delicious when marinated in this fruity sauce before being barbecued or oven-baked.

Serves 4

2½ cups cranberries
½ cup water
½ cup light brown sugar
2 tablespoons red wine vinegar
1 cup bottled hoisin sauce
3 pounds spareribs, trimmed

Combine the cranberries, water, sugar, and vinegar in a pan. Bring to simmering point, cover, and cook until the cranberries are very soft. Remove from the heat and stir in the hoisin sauce.

Spoon two-thirds of the cranberry mixture into a blender or food processor, and blend until smooth. Scrape into a pitcher and stir in the remaining cranberry mixture. Set aside until cold.

Spread out the ribs in a shallow dish which is large enough to hold them in a single layer. Pour the cranberry sauce over. Turn the ribs in the sauce until well coated, then cover and leave overnight in the refrigerator.

Next day, drain the ribs, reserving the marinade. Cook for 30 minutes, bone side down, on a greased grill over medium hot coals, then turn and continue to grill until cooked through.

Baste occasionally with the marinade. Pour the remaining marinade into a pan, bring to a boil, and cook for 2 minutes. Serve it as a sauce with the ribs.

tip
• If you prefer to cook the marinated ribs entirely in the oven, put them in a roasting pan, pour the marinade over, and cover with foil. Bake for 1 hour in an oven preheated to 400°F, then remove the foil, increase the oven temperature to 425°F, and cook for 20–30 minutes more.

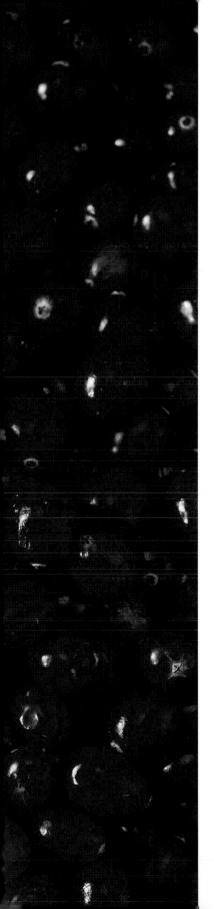

cranberry pie

This single-crust pie is very easy to make. The filling couldn't be simpler—a true celebration of the fine flavor of cranberries, accentuated with orange.

Serves 6

1½ cups all-purpose flour
pinch of salt
6 tablespoons butter
2 tablespoons sugar, plus extra for topping
1 egg
1–2 tablespoons milk
Filling
1¼ cups sugar
finely grated zest and juice of 1 orange
6 cups cranberries

Preheat the oven to 400°F. Make the pastry by mixing the flour and salt in a bowl. Cut in the butter, then rub it in until the mixture resembles breadcrumbs. Stir in the sugar, then add enough of the egg to make a dough. Wrap in plastic wrap and set aside in a cool place to rest for 30 minutes.

Make the filling. Mix the sugar, grated orange zest, and juice in a bowl. Mix well, then stir in the cranberries. Spoon into a deep pie dish with a capacity of 6 cups.

Roll out the pastry on a lightly floured surface to fit the top of the dish, adding an extra 1 inch all round. Cut off a 1-inch strip from around the edge. Dampen the rim of the pie dish lightly with water and stick the pastry strip in place. Add the milk to any remaining egg yolk, and brush a little of this on the pastry strip. Fit the pastry round on the pie. Press the edge to seal, then crimp it.

Decorate the top crust of the pie with pastry shapes, if you like. Cut one or two slits to allow steam to escape, then brush the pie crust with the remaining egg and milk mixture, and sprinkle it with sugar. Bake the pie just above the center of the oven for 25–35 minutes, until the pastry is golden and crisp. Sprinkle with more sugar just before serving.

berries

blueberry streusel tart

This has to be one of the best blueberry recipes there is. Under the almond streusel topping, the berries are beautifully succulent.

Serves 8

1¼ sticks butter, softened
3 tablespoons sugar
1 egg
2 cups all-purpose flour
¼ cup light cream
½ teaspoon pure vanilla extract
2 tablespoons finely ground almonds
2 tablespoons soft white bread crumbs
crème fraiche or plain yogurt, to serve

Filling

2½ pints blueberries
½ cup sugar
5 tablespoons soft white bread crumbs
3 tablespoons sliced almonds
2 tablespoons light brown sugar
½ teaspoon ground cinnamon

Preheat the oven to 400°F. Grease a 10½-inch quiche pan.

Make the pastry. Beat the butter and sugar together in a bowl. When the mixture is light and fluffy, beat in the egg with a little of the flour. Stir in the remaining flour alternately with the cream and vanilla extract, mixing to make a smooth, soft dough. Spoon the dough into the quiche pan, then use your fingers to gently ease it evenly over the base of the pan and up the sides.

Mix the ground almonds and bread crumbs together, and sprinkle the mixture evenly over the bottom of the pastry case.

Prepare the filling by mixing the blueberries with the sugar and half the bread crumbs. Spoon the mixture into the pastry case.

In a bowl, mix the remaining bread crumbs with the sliced almonds, brown sugar, and cinnamon. Scatter the mixture evenly over the blueberries.

Bake for 30 minutes, until the pastry is cooked and the streusel topping is golden. Serve warm or cold, with crème fraiche or yogurt.

variations

• You can use blackberries, currants, or gooseberries instead of blueberries. Adjust the sweetness as necessary, and add a little grated orange zest if you like.

blueberry muffins

Perfect for breakfast, brunch, or any other time of day. Do you really need an excuse to eat these tasty treats?

Makes 12–14

2 cups all-purpose flour
¼ teaspoon salt
1 tablespoon baking powder
½ cup sugar
1 cup blueberries
2 eggs
5 tablespoons light vegetable oil
½ cup milk

Preheat the oven to 400°F. Grease a 12–14 cup muffin pan.

Sift the flour, salt, and baking powder into a bowl. Stir in the sugar and blueberries.

In a separate bowl, whisk the eggs with the oil and milk.

Stir the liquids into the dry ingredients until just mixed. The mixture should be quite lumpy.

Fill the muffin pans so that each cup is two thirds full. Bake for 23–25 minutes, until the muffins are well risen and golden brown. Serve warm.

variations
• Using a zester, remove fine strips of orange or lime zest, and add these to the blueberries.
• Add some chopped nuts to the mixture. Almonds and pinenuts go well with blueberries.
• Split the cooked muffins and spread them with apple butter or lemon curd. Blueberry muffins are also good with a mild cheese, such as cream cheese or mascarpone.

tip
• The great thing about these muffins is that you can use frozen blueberries without thawing them first. They thaw and cook perfectly while the muffins are baking.

blueberries

black currant & apple pie

An old-fashioned pie, with a hint of spice. Blackberries, cranberries, or red gooseberries can be used instead of black currants and are equally delicious.

Serves 6–8

2½ cups all-purpose flour
½ cup confectioners' sugar
1¼ sticks butter
1 egg, plus 1 egg yolk
milk, for glazing
sugar for sprinkling

Filling

3 small tart apples
2½ pints black currants, stripped
   from stalks
1 cup sugar
1 tablespoon cornstarch
½ teaspoon ground allspice
½ teaspoon freshly grated nutmeg

Make the pastry. Sift the flour and confectioners' sugar into a bowl. Cut in the butter, then rub it in until the mixture looks like bread crumbs. Add the egg and egg yolk, and mix quickly to a rough dough. Wrap it and let it rest in a cool place (not the refrigerator) for 20 minutes.

Preheat the oven to 400°F. Cut the apples in quarters, remove the peel and cores, then slice thinly. Put the slices in a mixing bowl and add the black currants, sugar, cornstarch, allspice, and nutmeg. Toss gently to mix.

Roll out just less than half the pastry on a lightly floured surface, and line a 10-inch pie pan. Spoon the filling into the pan.

Roll out the remaining dough to a round 1 inch wider than the rim of the pan. Moisten the rim of the pie shell with water, then lift the pastry over the rolling pin and lay it on top of the filling. Press both pieces of pastry together on the rim and crimp the edge if you like.

Use any spare pastry to make decorative shapes for the top of the pie. Brush the top with milk, attach the shapes, then brush them with milk. Sprinkle the pastry with sugar.

Bake for 15 minutes, then lower the oven temperature to 350°F and bake for 20–25 minutes more. Serve warm.

black currants

70

red currant swirl

The contrast between the deep red of
the red currants and the white whorls
of sour cream is very effective.

Serves 4-6

1 cup sugar
2½ cups water
⅔ cup rosé or red wine
2 pints red currants, stripped from stalks
3 tablespoons arrowroot mixed to a paste with
    3 tablespoons water (see tip)
4–6 tablespoons sour cream

Put the sugar in a pan with the water. Heat, stirring,
until the sugar dissolves, then bring to a boil and
boil without stirring for 2–3 minutes. Stir in the wine.

Add the red currants to the wine syrup, lower the
heat, and poach them for about 10 minutes until
they are just tender.

Stir in the arrowroot paste. Bring to a boil, stirring all
the time until the mixture thickens. Let the mixture
cool, then chill in the refrigerator for several hours.
Serve in individual glass dessert dishes, swirling
a tablespoon of soured cream on the surface of
each portion.

tip

• Arrowroot is a better choice than cornstarch for
thickening the dessert because it gives a clear,
rather than a cloudy, result.

rhubarb

An ancient plant, rhubarb has a long and venerable history. It is believed to have originated in northern Asia, and was certainly known to the Chinese over 4,000 years ago. They valued the plant for its laxative qualities, and used it medicinally. Marco Polo is credited with introducing it to Italy. By the early 16th century, rhubarb was being grown in Britain, but as a cure for costiveness (constipation) or as an ornamental plant, rather than for food. During the reign of Queen Elizabeth I, and again during the First World War, rhubarb leaves were recommended as a substitute for cabbage or other leaf vegetables—but this was very dangerous advice, as they contain a chemical called oxalic acid. Large quantities could kill you, and even small amounts could make you feel very ill indeed.

It was during the 18th century that rhubarb finally began to be regarded as a fruit (even though it is technically a vegetable). Sweetened rhubarb started to be used in compotes, pies, and suet puddings. It was introduced into America in the early 19th century.

nutrition

Rhubarb is not particularly nutritious, but it is extremely low in calories. If you can tolerate it unsweetened (or sweetened artificially) it is a good choice for anyone trying to lose weight. It yields some vitamin B1, as well as calcium and potassium.

selection and storage

Rhubarb stalks should be firm and straight, with clear, strong color. Forced rhubarb is pale pink; main crop rhubarb is green or red and is often stringy. Avoid very thick stems; they can be tough. Use rhubarb as soon as possible after picking or buying it, as the stems rapidly become limp. If you must store it, wrap it well and keep it in the salad crisper of the refrigerator.

preparation

Remove and discard the leaves and the root end on each stem immediately. Wash the stems quickly in

cold water. If strings trail from the stems, pull them off. Slice the stems, cutting out any damaged or discolored areas.

cooking

Rhubarb is not generally eaten raw. Even young, forced rhubarb, which is naturally sweeter than older, red rhubarb, needs some sugar or honey to make it palatable and bring out the flavor. This can be added at the start of cooking, or at the end. You can cut down on the amount of sweetener needed by cooking rhubarb with another fruit, such as apple or banana, or by adding a dried fruit, such as apricot. Ginger goes very well with rhubarb. Rhubarb and ginger jam is especially tasty.

Rhubarb cooks quickly, and will become a mush if not watched closely. The stems yield plenty of juice, so it is only necessary to add a very small amount of water. The best way to cook rhubarb for a compote is to bake it in the oven just long enough for the slices to become tender. That way they will retain their shape and clear color. The fruit can be used in both savory and sweet dishes. Rhubarb sauce can be used instead of gooseberry sauce with mackerel, and also goes well with roast pork.

rhubarb compote

With its jewel-like color and intense flavor, rhubarb is the ideal fruit for a compote. Bake it in the oven, rather than on top, as this helps to stop it breaking up.

Serves 4

1½ pounds rhubarb, trimmed and sliced
¾ cup sugar
orange or lemon juice (see method)
dessert biscuits and plain yogurt,
    whipped cream, or
    crème fraiche, to serve

Preheat the oven to 350°F.

Toss the rhubarb and sugar in a baking dish. Cover with a lid or foil and bake 30 minutes, or until the rhubarb is tender, but retains its shape. Remove from the oven and leave to cool.

Using a slotted spoon, divide the pieces of rhubarb among four glass dishes. Taste the juices that remain in the bowl, and sharpen them with a little lemon or orange juice, if needed.

Spoon the juices over the rhubarb. Serve very cold, with dessert biscuits and yogurt, whipped cream, or crème fraiche.

variation

• Instead of serving the compote and cream separately, purée the rhubarb with a little of the juice, then fold it into sweetened, whipped cream.

rhubarb & ginger cake

Banish the blues with this warming and delicious cake. Whether you serve it with morning coffee or as a dessert, everyone will want a second helping.

Serves 6–8

1¾ cups all-purpose flour

2 teaspoons baking powder

¼ teaspoon salt

⅔ cup stoneground cornmeal

2 eggs

1 stick butter or margarine, softened

½ cup sugar

3 tablespoons milk

1 teaspoon pure vanilla extract

1 pound rhubarb, trimmed and cut into chunks

3 tablespoons drained, sliced preserved ginger
    in syrup

confectioners' sugar, for dusting

vanilla ice cream, plain yogurt, or crème fraiche,
    to serve

Topping

½ cup all-purpose flour

¼ cup ground hazelnuts

5 tablespoons light brown sugar

½ teaspoon ground ginger

½ stick butter, diced

Preheat the oven to 325°F. Grease a 9-inch springform cake pan. Sift the flour, baking powder, and salt into a bowl. Stir in the cornmeal. Beat the eggs until they are thick, pale, and fluffy.

In a separate bowl, beat the butter or margarine with the sugar until light and creamy. Fold in the dry ingredients alternately with the beaten eggs, milk, and vanilla extract. The mixture will be quite thick.

Spoon the mixture into the prepared pan, and level the surface. Arrange the rhubarb and preserved ginger on top. Make the topping by mixing the flour, ground hazelnuts, brown sugar, and ground ginger in a bowl, and rubbing in the butter until the mixture looks like bread crumbs. Sprinkle this over the rhubarb and preserved ginger, so that some pieces of rhubarb are quite thickly coated, while others are just dusted with the mixture.

Bake for 1–1¼ hours, by which time the cake base should be firm and the crumb topping golden. Test that the rhubarb is tender by gently inserting a skewer in one of the exposed pieces. Leave to stand in the pan for 5 minutes, then remove the sides and transfer the cake to a plate. Dust with confectioners' sugar and serve warm, with ice cream, plain yogurt, or crème fraiche.

tip

• Cornmeal adds an interesting texture and color to this cake, as well as helping the mixture to support the weight of the rhubarb and preserved ginger. Ground almonds could be used instead.

oranges & kumquats

All the world loves an orange. Golden globes of flavor, they fill our kitchens with color. We drink their juice, take them to school and work as a snack, and enjoy them in salads, sauces, and even soups.

There are two types of orange: the bitter orange, which is believed to have originated in India; and the sweet orange, which came from China. Bitter oranges are too sour to eat raw, but make marvelous marmalade. Arabs brought them to Spain, and they soon became popular throughout the Mediterranean. Sweet oranges came later, and were allegedly introduced to Portugal by the explorer Vasco da Gama.

Most of the world's oranges are grown within the citrus belt, which extends for 40 degrees north and south of the equator. There are over 2,000 varieties. The best known bitter orange is the Seville, while the most popular sweet varieties are the thin-skinned Valencia and the thicker-skinned Navel oranges. The flesh and juice of blood oranges are tinged with crimson. They are good for juicing, and in salads, especially with red onions and vinaigrette.

nutrition

Oranges are an excellent source of vitamin C and other antioxidants. They also contain some potassium and folate. If you peel an orange, then eat everything except the seeds, you will also benefit from the dietary fiber in the membranes and pulp.

selection and storage

Look for bright, shiny fruits that feel heavy for their size. Oranges should always feel firm, never shrivelled or soft, and should have a distinct citrus aroma. The color is not particularly relevant. If you buy ripe oranges from a street market in West Africa, for instance, the skins may be greenish, but the flesh will still taste delicious. Oranges are pretty robust and can be stored for up to two weeks at room temperature, if the room is not too humid. Turn them regularly, watching out for signs of mold.

preparation

Unlike the skin of most fruits, which is either eaten with the flesh or cut off and discarded, orange peel is valued for itself. It contains an aromatic oil that can be extracted if you rub sugar cubes over the skin; the cubes can then be crushed and used to flavor desserts or drinks. Strips of orange zest can be dried in the sun or in a low oven, then added to cakes, bakes, or meat dishes such as daubes or tagines. Remove the zest with a zester, if possible, as this is the best way to avoid including the bitter white pith that lies just below the surface. Alternatively, use a vegetable peeler. Check the pieces of zest and scrape off any of the bitter white pith with a small knife, then cut into thin strips. Either dry these or cook them in boiling water until they are soft, adding a little sugar if you like. Drain and use as a decoration or in sauces or desserts.The zest can also be grated. Always buy unwaxed fruit if you intend to use the peel.

For eating in the hand, you can either peel the orange neatly and break it into segments, or cut it in half and use your teeth to remove the flesh. This is satisfying, but tends to be messy.

To segment an orange perfectly, use a serrated knife to cut a thin slice off the top and bottom. Stand the fruit on a board and slice off the peel vertically in strips, taking care to remove the pith. Then, working over a bowl to catch the juices, cut between the membranes to release each segment. The segments can themselves be sliced thinly, but this is difficult to do unless you chill them first. For maximum vitamin C value, eat oranges as soon as possible after cutting.

kumquats

These small, bright orange ovals are not related to oranges, although they look like miniature versions of the fruit. Kumquats can be eaten whole, skin and all, but unless you have obtained a particularly sweet variety you may prefer to cook them. Poached in a light spiced syrup, kumquats are absolutely delicious. They keep well in the refrigerator, and can be used alongside savory dishes, such as baked ham, and desserts, such as parfaits or compotes.

blood oranges

carrot & orange soup

This has a strong, fresh citrus flavor. If you prefer the carrot to dominate, use only one orange.

Serves 6

¼ stick butter
1 onion, finely chopped
1 teaspoon ground coriander
½ teaspoon ground cumin
2 oranges
1½ pounds carrots, chopped
5 cups chicken stock
1 bay leaf
salt and pepper
carrot flowers (see tip), to garnish

Melt the butter in a heavy-bottomed pan and fry the onion over medium heat for 4–5 minutes until softened but not colored. Stir in the coriander and cumin and fry for 2 minutes more, stirring frequently.

Pare the zest from one of the oranges in a long strip, taking care not to take off any of the bitter white pith. Add the carrots to the pan, stirring to coat them in the butter and spice mixture, then pour in the stock. Add the bay leaf and pared orange zest. Bring to a boil, lower the heat, and simmer for about 40 minutes, until the carrots are very soft.

Remove the bay leaf and orange zest, then purée the soup in a blender or food processor. Pour it back into the pan.

Squeeze both oranges and add the juice to the soup. Stir in salt to taste, and add plenty of freshly ground black pepper. Heat through without boiling, and pour into heated bowls. Garnish each portion with a few carrot flowers.

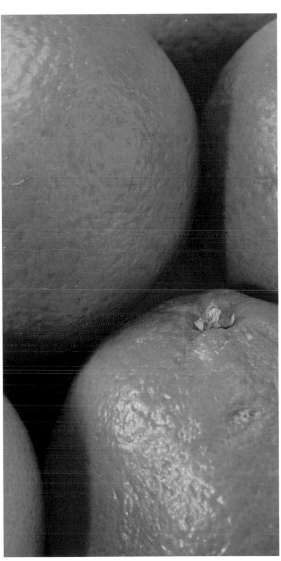

tip

• To make carrot flowers, peel a slender carrot, then use a channel knife to cut four furrows down its length, spacing them equally. When you slice the carrot in rings, each will look like a flower.

sea bass with orange & black bean sauce

Citrus and spice make an excellent sauce for a robust fish such as sea bass.
When first mixed the sauce will have a very strong flavor, but it mellows when
baked, and makes the fish taste wonderful.

Serves 6–8

3 pounds skinned sea bass fillets
3 oranges
2 tablespoons Thai fish sauce
5 tablespoons black bean sauce
2 garlic cloves, crushed
2 tablespoons light soy sauce
1 teaspoon grated fresh ginger
2 tablespoons sunflower oil
fresh chives, to garnish

Preheat the oven to 350°F. Arrange the pieces of
fish in a single layer in a large baking dish.

Peel two of the oranges, then chop the flesh,
discarding any seeds, and put it in a bowl. Stir in the
fish sauce, black bean sauce, garlic, light soy sauce,
ginger, and oil.

Pour the mixture over the fish, cover with foil and
bake for 15–20 minutes or until the fish is cooked,
and flakes easily when tested with a sharp knife.

Meanwhile, slice the third orange thinly. Put the
slices together in pairs. Make a cut from the rim to
the center of each pair. With the cut facing you, lift
one side up, at the same time turning the other side
down, to make a double twist. Make more double
twists in the same way.

Transfer the pieces of fish to a heated platter and
garnish with the orange double twists and whole

chives. Strain the cooking liquid into a pitcher.
Spoon a little over the fish and serve the rest
separately.

tip
• Fish cooked this way also tastes good cold. Serve
on a bed of shredded lettuce, with sliced cucumber
and green onions. Offer a little mayonnaise mixed
with sun-dried tomato paste and fresh lime juice, if
you like.

orange & avocado salad

Make this simple salad just before you intend to serve it, or the avocado will discolor.

Serves 4

1 bunch fresh baby spinach leaves
¼ small red cabbage
2 oranges
2 avocados
2 tablespoons lemon juice
1½ teaspoons sugar
½ teaspoon Dijon mustard
1 garlic clove, crushed
6 tablespoons light olive oil
salt and pepper

Rinse and dry the spinach leaves. Tear any large leaves into pieces and put them in a salad bowl.

Cut out and discard the core from the red cabbage, then shred the cabbage very finely. Toss the shreds with the spinach leaves. Peel the oranges, taking care to remove all the bitter white pith. Working over a bowl, hold an orange in your hand and carefully cut between the membranes to release each orange segment in turn. When all the segments have dropped into the bowl, with any juice that has been released, squeeze the orange pulp over the bowl to extract the rest of the juice. Repeat with the second orange.

Cut the avocados in half and lift out the pits. Remove the peel from each half, then slice the avocado halves lengthwise. Add the pieces to the orange segments.

Make a dressing by whisking the lemon juice, sugar, mustard, and garlic in a small bowl. Whisk in the oil,

add salt and pepper to taste, then whisk in the orange juice from the bowl containing the avocados and oranges.

Using a slotted spoon, add three-quarters of the orange segments and avocado slices to the spinach and red cabbage mixture. Pour over about 2 tablespoons of the dressing and toss lightly.

Arrange the remaining orange segments and pieces of avocado on top of the salad. Serve immediately. The remaining dressing can be poured into a pitcher and offered at the table, or saved for another salad.

variations
• Crumble fried bacon over the salad.
• Cut strips of smoked salmon and thread them, accordion-fashion, on toothpicks. Serve the salad on individual plates and set two miniature smoked salmon brochettes on each one.
• Serve the salad with cooked shrimp or rollmop herrings.

upside-down kumquat cake

Some people know this as topsy-turvy cake, but whatever you call it, it is a real family favorite.

kumquats

Serves 6–8

1½ sticks butter
6 ounces kumquats
4 tablespoons light brown sugar
½ cup sugar
1 cup all-purpose flour
1 teaspoon baking powder
¼ teaspoon salt
2 eggs, beaten
½ teaspoon natural vanilla extract
cream or ice cream, to serve

Preheat the oven to 350°F. Melt ½ stick butter in the bottom of an 8-inch square cake pan. Dip a pastry brush in the butter and lightly grease the sides of the pan.

Bring a small pan of water to a boil, add the kumquats, and cook for 10 minutes. Drain well. When cold enough to handle, cut the kumquats in half lengthwise and scoop out the seeds. Sprinkle the brown sugar evenly over the bottom of the pan. Arrange the kumquat halves, cut sides down, in a pattern over the brown sugar.

Beat the remaining butter and sugar together in a bowl. Add the flour, baking powder, and salt alternately with the eggs. Beat in the vanilla extract.

Drop the mixture in spoonfuls into the pan, then gently smooth it together, taking care not to disturb the kumquats. Level the surface. Bake in the center of the oven for 30–35 minutes, or until a skewer inserted in the cake comes out clean.

Let the cake cool in the pan for 5 minutes, then invert it on a plate. Serve warm, with cream or ice cream.

variations

• To intensify the orange flavor, stir in 2 tablespoons of orange marmalade.
• Try this with other types of fruit, too. Sliced fresh peaches or nectarines work well, but you can also use drained canned fruit, such as pineapple rings.

spiced kumquats

Kumquats can be eaten raw, but not everyone likes them that way. Cooked in spiced syrup, however, kumquats become tender and sweet, and taste wonderful with both sweet and savory dishes.

Makes 1 large jar

1 cup sugar
1½ cups water
1 cinnamon stick
6 cloves
1½ pounds kumquats
2 tablespoons orange-flavored liqueur (optional)

Put the sugar in a pan with the water. Heat, stirring, until the sugar dissolves, then add the cinnamon stick and cloves. Bring to a boil, and boil without stirring for 2–3 minutes.

Wash the kumquats. Pinch off the little green tops, then add the fruit to the syrup.

Lower the heat, cover the pan, and simmer the fruit for 20–30 minutes or until the kumquats are tender and look a bit deflated.

Remove the pan from the heat, and let stand for 5 minutes.

Ladle the fruit into warm, sterilized jars, pressing it down well. If using the liqueur, stir it into the kumquat syrup, then spoon it over the fruit.

Leave until cold, then top up the level of the syrup if necessary. Close the jars and store them in the refrigerator.

tips
• Spiced kumquats are very good with hot baked ham, or with slices of cold ham, and come in very handy at Christmas time.
• They are also good over ice cream.
• A jar of spiced kumquats makes a lovely gift.

kumquats

tangerines, mandarins, & satsumas

It can be confusing to go to your local market and find displays of bright orange fruit that all look more or less the same, but which have completely different names. Tangerine, mandarin, satsuma, tangelos, orantiques—what precisely are they?

All these fruits are related to the sweet orange, and like the sweet orange, they originated in China. The loose, easy-to-peel skins and simple-to-separate segments made them popular wherever they were cultivated. The name tangerine is supposed to have originated from a variety grown in Tangiers, Morocco, while the satsuma was developed in the province of the same name on the island of Kyushu in Japan. The clementine originated in India, but is now widely grown in North Africa, Spain, and Israel. Tangelos are a cross between mandarins and grapefruit, while orantiques are orange-tangerine crosses.

The name mandarin celebrates the Chinese connection. In China, the fruit are traditionally cut from the trees so that the dark, glossy leaves provide a contrast to the beautiful orange fruit.

Mandarins are used to decorate food for the Chinese New Year celebrations. In England, tangerines are traditionally inserted into the toes of Christmas stockings before they are filled with small gifts.

nutrition
All these fruits have a high vitamin C content, although not as high as oranges. They are rich in beta-carotene, which the body converts to vitamin A, and have a high fiber content.

selection and storage
Some fruits have soft, saggy skin, whereas others are taut. The skin should be bright orange all over, with no trace of damage or mold. Keep tangerines in the fruit bowl only if you are likely to eat them within a day or two; otherwise store them in the refrigerator. Puffy fruit—the types with saggy skins—tend to keep less well than cling-skinned varieties.

preparation
All tangerines are very easy to peel, but are quite fibrous. It can be difficult to remove the string-like pith from the segments. It tends to stick to your fingers as you try to pick it off. A much easier way to remove it is to dip the segments briefly in boiling

water, then scrape off the pith with a small, sharp knife. Tangerines are very handy fruits, ideal for lunch boxes. Segments can be used in fruit salads, or half-dipped in melted chocolate, then left to set on wax paper for an after-dinner treat. Tangerine peel dries beautifully. Like dried orange peel, it is often added to curries and stews, or used in cakes and bakes. Grating is not an easy option, as it is very difficult to avoid including the bitter pith. These fruits are not as juicy as oranges, but the juice tends to have a more concentrated flavor.

cooking

Fresh fruits are seldom cooked (the flavor is spoiled if they are boiled), but segments can be warmed through in fruit sauces. Like most members of the citrus family, tangerines and their relatives go well with poultry, especially duck and game birds.

mandarin segments

Canned mandarin segments are a familiar ingredient. Children like them in Jell-O, and they can be added to fruit salads when fresh fruit isn't available. What are less well known, but just as useful, are canned whole mandarins. Packed in fruit juice—often apple juice—they are perfect for making an impressive dessert in next to no time. Drain the contents of 2 cans of whole mandarins, then pour half the juice into a pan and warm it gently. In another pan, heat 1 cup sugar with ¼ cup water, stirring until the sugar has dissolved. Stop stirring and boil the syrup until it turns golden brown, then remove it from the heat and carefully pour in the warm apple juice. Heat gently to dissolve the caramel, then set aside until cold. Pour the cold caramel sauce over the whole mandarins in a glass serving dish and chill thoroughly before serving.

lemons & limes

Where would we be without lemons and limes? So much of what we eat and drink is improved by their astringency, from the pre-dinner gin and tonic or mineral water to the refreshing sorbet with which we conclude our meal. Avgolemono, the Greek chicken and lemon soup, lemon chicken, fish with a twist, lemon soufflé—the list goes on and on. At one time, the use of lemons was much more widespread than that of limes, but now that more and more of us are discovering just how delicious is the flavor of these dark green citrus fruits, limes are becoming just as popular. They are especially good with West Indian and South American food, and are famously used to make ceviche, where raw fish fillets are "cooked" simply by being marinated in lime juice.

Limes and lemons originated in India, and are now cultivated throughout the world. Limes are largely limited to tropical and sub-tropical areas, whereas lemons are more widely distributed. Unlike limes, they grow well in Mediterranean countries and in North Africa.

tahitian and key limes
Lemons are seldom sold by variety, but you will find more than one sort of lime at the market. Tahitian limes are almost as big as lemons. They are grown in California, along with Bearss and Persian limes. A smaller variety is the Key lime, which is cultivated in Florida and the West Indies. Key limes are more astringent that other types, so if they are specified in a recipe and you cannot locate any, add a little lemon juice to create the correct flavor.

nutrition
Both lemons and limes are rich in vitamin C, although lemons yield almost twice as much as limes. They also have antiseptic properties, so the juice is recommended for anyone with a sore throat.

selection and storage
When buying lemons or limes, look for fruit that is

brightly colored and heavy for its size. Avoid lemons that are tinged with green, as this can mean that the flesh inside is sour. If you are buying lemons for juicing, choose smooth-skinned fruit; if it is important that the rind can be grated easily, rough-skinned lemons are better. Both lemons and limes are reasonably robust, but limes have thinner skins, so must be handled with more care. As with all citrus, lemons and limes can be kept at room temperature for short periods, but if you want them to last, keep them in the refrigerator. Don't store them for too long or their fragrance will fade.

preparation

Lemon and lime zest are both widely used for flavoring. Use a zester if possible, to remove the zest in strips, and make sure you don't include any of the bitter white pith. When grating lemons or limes, work over a piece of foil. Having grated the fruit, brush down the grater with a pastry brush, then use the brush to sweep the grated rind off the foil and into your bowl. Both lemons and limes will yield more juice if you roll them on a hard surface or warm them gently first. Use a lemon squeezer for juicing if you like, but for small quantities, a reamer (a wooden implement with a ridged bulbous end) works even better. Cut the fruit in half, then use the reamer to press the juice into a strainer set over a bowl to catch the seeds. Water acidulated with lemon juice will prevent fruits such as apples and pears from discoloring.

For segmenting lemons, see the chapter on oranges.

cooking

Lemons and limes are seldom cooked entirely on their own, but are used for accent with other foods. A whole lemon, pierced with a skewer and inserted in the cavity of a chicken before roasting, will impart a delicious flavor.

roast lemon chicken

Everyone loves roast chicken, and this version is especially good, with its lemon and whole wheat stuffing.

Serves 4–6

2 lemons
1 stick butter
1 small onion, finely chopped
1 celery stalk, chopped
½ cup sliced almonds
2 cups fresh whole wheat bread crumbs
salt and pepper
lemon wedges, to serve

Preheat the oven to 400°F. Grate both lemons and squeeze the juice from one of them. Save the squeezed lemon halves. Melt 2 tablespoons of the butter in a small pan and fry the onion with the celery until softened. Remove from the heat and stir in the sliced almonds, bread crumbs, and half the grated lemon zest. Stir in enough of the lemon juice (2–3 tablespoons) to bind the mixture, then season with salt and pepper.

Stuff the chicken loosely with the bread crumb mixture, using one or both of the squeezed lemon halves to hold the stuffing in place.

Gently ease the skin away from the chicken breast, so that the flesh underneath is accessible. Take care not to break the skin. Cream the remaining butter with the remaining lemon zest. Gradually work in 1 tablespoon of the remaining lemon juice. Spoon half the mixture between the breast and the skin, smoothing it gently to coat evenly. Rub the remaining lemon butter over the skin.

Roast the chicken for 2 hours, basting occasionally with the pan juices. When the chicken is cooked, the legs will move freely, and any juices that emerge when the thickest part of the thigh is pierced with a skewer will be clear. Lift the chicken onto a heated platter, garnish with the lemon wedges, and serve.

tip

• The lemon butter that is placed over the breast of the chicken before roasting helps to keep the meat beautifully moist, and gives it a wonderful flavor; but if you want to reduce the fat content of this dish, leave it out.

lemon soufflé

Like all cold dessert soufflés, this is baked in a dish with a paper collar, so that when the collar is removed, it looks as though the soufflé has risen above the level of the dish, much as a baked soufflé might do.

Serves 4–6

This recipe contains raw egg, see note on page 142.

2 large lemons
1 tablespoon powdered gelatin
4 eggs, separated
½ cup sugar
1½ cups heavy cream
toasted almonds and whipped cream, to decorate

Prepare a 3-cup soufflé dish. Measure the height of the dish and add 3 inches. Cut a piece of parchment paper twice as wide as this measurement and long enough to go around the dish with an overlap. Fold the paper in half lengthwise and tie it around the dish, using adhesive tape to hold it in place. Set the dish on a large plate.

Grate the zest from one of the lemons and set aside. Squeeze both lemons and measure the juice. You will need ¾ cup; make up the amount with water, if necessary. Heat the lemon juice in a small pan and stir in the gelatin until dissolved.

Whisk the egg yolks with the lemon zest and sugar until pale and creamy. Gradually add the gelatin mixture, whisking all the time. Continue to whisk for 3–4 minutes to incorporate as much air as possible. Set aside in a cool place until the mixture is at the point of setting.

Whisk the egg whites until they are stiff. In a separate bowl, whip the cream until soft peaks form. Fold first the cream and then the egg whites into the lemon mixture. Pour into the prepared soufflé dish. The mixture will fill the dish and rise above it, supported by the paper.

Chill in the refrigerator for 3–4 hours until set. Run a knife around the top of the soufflé, where it joins the paper collar, then undo the collar and peel it away.

Press the toasted almonds onto the exposed sides of the soufflé, and decorate the top with rosettes of whipped cream, decorating each rosette with a toasted almond. Serve at once.

tip

• Use a clean whisk to beat the egg whites, as any trace of egg yolk will prevent the whites from beating to a good height.

lemon & lime love cake

This loaf is known by lots of different names, but Love Cake is particularly appropriate, since everyone who tries it finds it impossible to resist.

Makes 1 loaf

1 lemon
1 lime
1½ cups all-purpose flour
2 teaspoons baking powder
¼ teaspoon salt
¾ stick butter, softened
1¼ cups sugar
2 eggs, lightly beaten
½ cup milk

Preheat the oven to 325°F. Grease and flour a 9-x-5-inch loaf tin. Grate the zest from the lemon and lime and set it aside. Cut both fruits in half and squeeze one lemon half and all the lime. Mix the flour, baking powder, and salt together.

Put the butter in a mixing bowl and add ¾ cup of the sugar. Beat the mixture until it is pale and creamy, then gradually beat in the eggs, adding a little of the flour if the mixture shows signs of curdling. Gradually add the remaining flour mixture, alternately with the milk, beating well after each addition. Stir in the lemon and lime zest.

Spoon the mixture into the prepared loaf pan and level the surface. Bake for 40–50 minutes, until a skewer inserted in the center of the cake comes out clean. Mix the citrus juice with the remaining sugar. Invert the loaf onto a wire rack, then turn it right way up again.

Put a tray underneath the rack. Immediately spoon the sugar mixture over the top of the loaf, letting it run down the sides slightly. Leave to cool before slicing.

variations
• Add ½ cup chopped pecans with the lemon zest.
• Instead of using both lemon and lime, use the grated zest and juice of 1 lemon or 2 limes.

key lime pie

Smaller and more tart than other varieties, key limes give this pie a wonderful tang. They can be difficult to locate, however, so this recipe uses regular limes and a little lemon juice.

Serves 8

2 cups graham cracker crumbs
¾ stick butter, melted

Filling

5 egg yolks plus 3 egg whites
1 14-ounce can sweetened condensed milk
½ cup fresh lime juice
4–6 tablespoons fresh lemon juice
grated zest of 1 lime
5 tablespoons sugar

Mix the graham cracker crumbs with the melted butter. Tip the mixture into a 10-inch quiche pan and spread it evenly. Rest a small saucepan on the crumbs and use it to compress them. Use a circular movement, then take the pan right to the rim of the quiche pan so that it forces the crumbs up the sides of the tin to make an even biscuit case. Chill while you make the filling.

Preheat the oven to 350°F. In a large bowl, whisk the egg yolks until they are pale and thick. Gradually whisk in the condensed milk, then the lime juice and 4 tablespoons of the lemon juice. The mixture must be quite tart, so add the extra lemon juice if necessary. Fold in the lime rind. Spoon the mixture evenly into the biscuit case.

Whisk the egg whites until soft peaks form, then gradually whisk in the sugar until the mixture is stiff. Spoon it carefully over the filling, taking it right to the edge of the crumb crust. Swirl it attractively.

Bake for 15 minutes, until the meringue topping is lightly browned on the peaks and around the edges. Set the pan on a wire rack to cool for 1 hour, then chill the pie for 2–3 hours before serving.

limes

91

grapefruit

Grapefruit owes its origin to the pomelo, an even larger yellow fruit that comes from Polynesia. The pomelo's alternative name, shaddock, stems from the fact that a certain Captain Shaddock is credited with introducing it to Jamaica, where it did extremely well. The pomelo was crossed with the sweet orange, and the grapefruit was the result. Ugli fruit is a cross between a grapefruit and a tangerine (possibly with orange, too). It originated in Jamaica, but was named by the Canadians, who on first receiving a shipment, remarked unfavorably on its appearance. Far from being offended by the name, the Jamaicans patented it. Ugli fruit may not look good, but it has a delicious flavor. It has pinkish flesh and is sweeter than grapefruit.

Grapefruit is delightfully refreshing. It makes the perfect start to the day, is an excellent appetizer, and is also popular as a palate cleanser between courses. Its clean taste cuts through any excessive creaminess, so it is perfect on a cheesecake. The segments are superb in salads, and they not only go well with other fruits, but are also very good with ham, crisp bacon, and shellfish.

Grapefruit can be yellow (classified white), pink, red, or ruby. The sweetie is a cross between a grapefruit and a pomelo. It is green skinned and, as the name suggests, has lovely sweet flesh.

nutrition
All these citrus fruits are good sources of vitamin C and other antioxidants, including beta-carotene. They are high in dietary fiber.

selection and storage
Choose firm, heavy fruit. The skin can be thin or thick, but avoid fruit that looks flabby or whose skin looks coarse. In the refrigerator, grapefruit will keep for 7–10 days.

preparation
To serve grapefruit halves as a starter, cut the fruit in half around the equator. Using a curved serrated knife made for the purpose, cut around the rim and between the membranes to loosen each half

segment of flesh, then cut out the white core in the center. Sprinkle the fruit with a little sugar, if you like, although this should not be necessary with pink grapefruit or sweeties. To segment a grapefruit, follow the advice for oranges. Ugli fruit is prepared in the same way, but a pomelo, which has a thick layer of pith, needs a different technique. Slice off the top and bottom of the fruit, then use a sharp knife to score the skin deeply, cutting from top to bottom so that you effectively mark the skin into eight equal divisions. Peel back each division in turn and cut away the thick pith to reveal the fruit segments.

Grapefruit halves, hollowed out and scraped clean, can be used to hold jelly. They look particularly good if you vandyke the fruit (use a zig-zag cut when halving it to create an interesting edging).

Grapefruit juice is delicious as a drink, and also as the basis for a salad dressing. The peel can be candied.

cooking

Warming grapefruit really brings out the flavor. Grilled grapefruit was once such a common starter that it became rather hackneyed, but it is still a very good method of serving this tasty fruit. Just cut the grapefruit in half, loosen the half segments as described above, then sprinkle the surface with brown sugar. Grill until the sugar caramelizes. You can douse the fruit in rum or sprinkle it with Angostura bitters first, if you like, or use honey instead of brown sugar.

grapefruit & kiwi salad

The tang of fresh grapefruit provides a perfect counterpoint to the sweetness of the kiwi fruit and grapes in this refreshing summer salad.

Serves 4

2 grapefruits
1 teaspoon sugar
6 tablespoons light olive oil
1 teaspoon soy sauce
4 kiwi fruit
1 head chicory (curly endive)
1 small bunch red grapes

Peel the grapefruit, taking care to remove all the bitter white pith. Working over a bowl to catch the juices, carefully cut between the membranes to release each grapefruit segment in turn. When all the segments have been released, squeeze the grapefruit pulp over the bowl to extract the rest of the juice. Scoop out any seeds that have fallen into the bowl.

Pour 2 tablespoons of the grapefruit juice into a small bowl, and whisk in the sugar until completely dissolved. Gradually whisk in the olive oil, then the soy sauce.

Peel the kiwi fruit and cut them in quarters. Mix the kiwi fruit and grapefruit segments in a salad bowl. Add the dressing and toss lightly. If you have time, set the mixture aside for about an hour to allow all the flavors to develop. Drain the fruit, reserving the dressing.

Wash the chicory (curly endive) leaves and dry them thoroughly. Tear them into bite-size pieces. Put them in a bowl and toss with 1 tablespoon of the grapefruit dressing, then spread them out evenly on a large salad platter.

Arrange the grapefruit segments and kiwi quarters alternately in one or more concentric circles on top of the salad leaves. Strip the grapes from the stems and pile them in the center. Serve at once, offering the remaining dressing in a small pitcher.

pink grapefruit cheesecake

This is an unusual cheesecake in that the crumb mixture is sprinkled on top of the filling, and the cheesecake is later inverted, so that it ends up on the bottom. This means that you can keep the amount of crumbs to a minimum, so it will appeal to those who do not like a thick, soggy crust.

Serves 6–8
This recipe contains raw egg, see note on page 142

3 pink grapefruit
¼ cup sugar
1 tablespoon powdered gelatin
1 egg, separated
1 14-ounce package cream cheese
⅔ cup heavy cream
½ cup cookie crumbs
whipped cream, to decorate

Line a 7-inch cake pan with parchment or wax paper. Finely grate the zest from 1 grapefruit, then squeeze it and measure ⅔ cup of the juice. Peel the remaining grapefruits and cut into neat segments.

Put half the sugar into a heatproof bowl and add the gelatin. Stir in the egg yolk and measured grapefruit juice. Place over gently simmering water and stir until the gelatin has dissolved. Remove from the heat and leave to cool.

Beat the cream cheese until softened, then beat in the gelatin mixture and 1 teaspoon of the grated

grapefruit zest. Chop 5 grapefruit segments and stir them into the mixture. Put the remaining segments in a covered bowl in the refrigerator.

Whip the cream. Whisk the egg white until it forms soft peaks, then add the remaining sugar and whisk until stiff. Fold first the cream, then the egg white, into the grapefruit mixture, then spoon it into the prepared pan. Level the surface.

Sprinkle the cookie crumbs evenly over the surface and press them down gently. Chill the cheesecake overnight in the refrigerator. When ready to serve, use a knife to ease the cheesecake away from the pan, if necessary, then invert onto a serving platter so that the thin crumb crust is underneath. Lift off the paper. Decorate the cheesecake with the whipped cream and the reserved grapefruit segments.

tip
• This gives a very thin crumb crust. Use more crumbs if you like, but remember that the cream mixture must bear their weight. Any cookies can be used, but gingersnaps are particularly delicious.

grapefruit

95

kiwi fruit

This very popular fruit was once called the Chinese gooseberry, but it is now so well known by the name given to it by the New Zealand growers who put it on the map that most people have forgotten it was ever known as anything else. Kiwi fruit—or kiwis—look like furry brown eggs and are not very prepossessing until you slice one open and reveal the pretty green interior patterned with tiny black seeds. They are often used as a garnish, either cut in slices or wedges, and taste good in fruit salads, sorbets, or on cheesecakes or pavlovas. Their slightly acidic flavor goes well with meat and poultry.

nutrition
Kiwi fruit contain vitamin C, dietary fiber and potassium.

selection and storage
Choose kiwi fruit that are firm and plump, with no bruising. The fruit should yield a little when gently squeezed. Hard kiwis will ripen if you put them in a brown paper bag with an apple or a banana, but if they are ripe already, avoid contact with other fruits as they rapidly become over-ripe.

preparation
To eat the fruit as a snack, treat it like a boiled egg. Just slice the top off and scoop out the center with a spoon, leaving the inedible furry brown skin behind. For a dinner party dessert that will amuse and delight your guests, serve kiwis in egg cups with fingers of brioche topped with chocolate spread. If you are using kiwi fruit in a composite dish, remove the skin with a sharp knife or vegetable peeler, then slice each fruit into rounds. Alternatively, cut it lengthwise in quarters or sixths. If you cut the fruit into segments, you may want to cut away some of the white core from each piece, but this is not really necessary.

cooking
Kiwi fruit tastes wonderful when warmed through, but don't attempt to cook it, or the pieces will disintegrate and turn an unattractive shade of khaki. The fruit is excellent tossed into a stir-fry, or served alongside fried fish.

kiwi fruit

96

beef & kiwi fruit stir-fry

Kiwi fruit tastes great in savory dishes as well as sweet ones. Warming it brings out the flavor, and it provides a great contrast to the spicy steak.

Serves 4

1 pound sirloin or strip steak

2 tablespoons sesame oil

2 tablespoons sweet chili sauce

1 tablespoon sunflower oil

½ garlic clove, peeled

1 slice fresh ginger

3 green onions, sliced diagonally

2 celery stalks, sliced diagonally

8 fresh or drained canned baby corn cobs

1 tablespoon soy sauce

1 tablespoon rice wine or sherry

3 kiwi fruit, peeled and quartered lengthwise

Trim the excess fat from the steak and cut it into thin strips. Mix the sesame oil and chili sauce in a shallow dish. Add the steak strips and toss them until they are well coated. Cover the bowl and set it aside for 30 minutes.

Heat a wok. Drizzle the sunflower oil around the top, just below the rim, so that it runs down to coat the surface. When the oil is hot, add the garlic and ginger and stir-fry for about 30 seconds, keeping them moving so that they do not scorch. Lift the garlic and ginger out and discard them.

Add the green onions, sliced celery and corn cobs to the flavored oil, and stir-fry for 3 minutes. Lift the vegetables out with a slotted spoon and put them in a bowl.

Drain the steak strips, reserving the marinade. Add them to the wok and stir-fry for 2–3 minutes, then return the vegetables to the wok and add the soy sauce and rice wine or sherry, with enough of the marinade to moisten. Toss over the heat for 1 minute, then add the kiwi fruit and cook for 1 minute more, tossing gently to avoid breaking up the fruit. Serve at once.

tip

• Serve this with noodles or a mixture of jasmine rice and wild rice.

kiwi fruit

pineapples

The pineapple has been called the king of fruits, and with its crown of spiky green leaves and imposing shape, it is certainly impressive. The fruit comes from Central and South America, where it was cultivated long before Columbus sailed the ocean and brought it home to Spain. Europeans had never seen anything like it before, and among the aristocracy it soon became a matter of prestige to have tasted a pineapple, or, better still, to have one growing in your own hothouse.

Many of the world's pineapples still come from the Caribbean region, but they are also exported from tropical Africa and Hawaii. There are several different varieties, some with golden-yellow skin, others that remain green-tinged even when ripe. The flesh, too, can range from pale yellow to a honeyed gold. Small pineapples—sometimes called baby pineapples, although they are fully mature—are also cultivated.

selection and storage

For a long time, buying a pineapple was a bit of a hit and miss affair, unless you were lucky enough to live in a country where they were picked straight off the plants. Although the fruit traveled well, it tended to be picked before it was fully ripe, and the flavor could be disappointing. Improved transportation, together with the development of new varieties that are particularly sweet and luscious, means that quality fruit is now easy to come by.

When buying a pineapple, your nose is the best judge. The fruit should have a perceptible perfume. The crown of leaves—the plume—should look glossy and healthy, not dull and faded, and there should be no sign of damage or bruising on the scaly looking skin.

Ripe pineapple should be prepared and eaten as soon as possible after purchase. If you cannot use it immediately, slice it and store the slices in a sealed tub in the refrigerator.

preparation

There are several ways of preparing a pineapple. If you want to end up with pineapple slices or wedges, to serve fresh or in fritters, cut off the leafy plume, then remove the skin in wide strips, cutting diagonally so as to remove the black "eyes" that are found on the outside of the flesh. These have an unpleasant, peppery flavor; any that remain must be gouged out with the tip of a vegetable peeler. After peeling the pineapple, cut the cylinder of flesh that remains into slices, then remove the central core from each slice with an apple corer.

If you want to hollow out the pineapple and stand it upright as a container for a fruit salad or salsa, start by cutting off the leafy top and setting this aside for the lid. Remove the flesh from inside, chop it and mix it with such extra ingredients as are required, then pile the mixture back into the shell and replace the lid.

Halved pineapples also make good receptacles when hollowed. Leave the plume of leaves on for the most dramatic effect, cutting it in half with the body of the pineapple. Another possibility is to make pineapple boats. Cut the pineapple in quarters, again including the plume, then cut out the core from each one. Without removing the flesh entirely, slide a knife between flesh and skin on each wedge and use a sawing action to separate the two, then cut the flesh widthwise in bars. Still leaving the flesh on the skin, push alternate bars of pineapple flesh to the left and right.

cooking

Pineapple wedges or slices are delicious grilled with a honey, brown sugar, or marmalade glaze. Pineapple goes well with beef, pork, and poultry, and is good in a stir-fry. Never use fresh pineapple in a gelatin-based dish or with fresh cream, as the fruit contains an enzyme which will prevent gelatin from setting, and will curdle cream. The enzyme is, however, a good tenderizer, so a pineapple marinade is an excellent idea. But don't marinate meat for too long, or the flesh will become mushy.

fresh pineapple salsa

Choose a pineapple with a good leafy plume, to make a container for the salsa.

1 small pineapple, with leafy top intact
1 small red onion or 4 green onions, finely chopped
1 red bell pepper, finely diced
3 tablespoons fresh lime juice
1 tablespoon extra virgin olive oil
salt and pepper
2 tablespoons chopped fresh mint

Cut a thin slice off the bottom of the pineapple, if necessary, so that it will stand upright. Slice off the leafy top, with a little of the pineapple beneath it, to make a lid. Using a sharp knife, cut out the flesh from the center of the pineapple, leaving a shell that is at least ½ inch thick. Set the pineapple shell and lid aside.

Dice the pineapple flesh finely. You should not find any "eyes" on the pineapple pieces, but if you do, cut them off. Put the diced pineapple in a bowl and add the onion or green onions and red bell pepper. Mix lightly.

Pour the lime juice into a small bowl and whisk in the olive oil. Season with plenty of salt and pepper, and stir in the mint. Pour the dressing over the pineapple mixture and toss well. Let stand for at least 1 hour to allow the flavors to blend.

Just before serving, stand the pineapple shell on a plate and spoon the salsa into the center. Replace the leafy top.

tip

• You don't have to use the pineapple shell if you don't want to. Simply serve the salsa in a bowl, and garnish with a sprig of fresh mint. An alternative way of serving is to cut the pineapple in half lengthwise, slicing right through the leafy top, then remove the flesh from each half. This gives you two containers for salsa, ideal for a buffet table.

glazed pineapple boats with sabayon

These taste as good as they look, the rum glaze giving a golden topping to the golden pineapple flesh. The sabayon sauce is the ideal accompaniment, but you can serve the pineapple boats on their own.

Serves 8

2 ripe sweet pineapples, with leafy tops intact
8 tablespoons marmalade
2 tablespoons butter
2 tablespoons dark rum
Sauce
5 egg yolks
5 tablespoons sugar
½ cup sweet white wine

Using a sharp knife, cut each pineapple in half lengthwise, cutting right through the leafy top. Cut each piece in half lengthwise twice more, to give eight wedges. Cut out the core at the center of each wedge and throw it away.

Starting at the leafy end of each wedge, slide a sharp knife between the pineapple flesh and the skin, then cut the flesh away in a solid wedge, using a sawing action. Leaving the flesh on the skin, cut it widthwise into slices. Place the pineapple boats in a broiler pan.

Melt the marmalade with the butter in a small pan, then stir in the rum. Spoon the mixture over the pineapple flesh.

Preheat the broiler. Place the pineapple boats 3–4 inches below the heat source while you make the sauce.

Put the egg yolks in the top of a double boiler. Whisk in the sugar. Set the pan over simmering water and add the wine, a little at a time, whisking constantly until the mixture starts to thicken. Keep the heat in the lower pan well below boiling point; if the sabayon sauce gets too hot it may separate. Keep whisking until the sauce is thick enough to leave a trail on the surface when the whisk is lifted. This can take up to 15 minutes.

When the sabayon sauce begins to thicken, move the pineapple nearer the heat source, so that the marmalade topping begins to caramelize.

As soon as the sabayon is ready, remove the pineapple wedges from the heat and place them on individual plates. Push alternate pieces of pineapple on each wedge to left and right, to reinforce the image of a boat. Spoon a little of the sabayon sauce onto each plate and serve the rest separately.

tip

• You can use any flavor of marmalade. The classic Seville orange marmalade works very well, but this recipe is also delicious with ginger or lime marmalade. Alternatively, use apricot or peach jam.

pineapples

mangoes

If you don't like mangoes, you probably haven't had a good one. At their best, mangoes taste superb, with subtly scented orange flesh that really does melt in the mouth. Their sweetness is such that they do not need sugar. In fact, they benefit from a squeeze of fresh lime juice.

Mangoes originally came from India, Indonesia, and Malaysia, and have been cultivated for centuries. The most familiar fruit is kidney-shaped, but there are also round and oval varieties. The thin skin is usually a delicate peachy-orange color, often blushed with red, but it can be yellow or even green. Some mangoes are more fibrous than others. The West African ruby mango is particularly so. The best way to enjoy it is to cut a small hole in the skin and suck out the plentiful juice.

Mangoes are best eaten fresh, in the hand, but they also make excellent ice cream and sorbet. A mango sauce or coulis is the ideal accompaniment for other fruit, or for a chocolate terrine. Green mangoes are used to make chutney.

nutrition

A source of vitamin C and beta-carotene, mangoes also contribute fiber and potassium to the diet.

selection and storage

Ripe fruit will be soft to the touch and have a fragrant aroma. This is what you should buy if you intend to eat the fruit right away. Firm fruit (provided it is not too hard) will ripen in a few days, if kept at cool room temperature. To hasten the process, put the fruit in a paper bag with an apple or banana. Ripe mangoes can be kept in the refrigerator for up to two days.

preparation

Inside every mango is a large pit, with fibers that connect it to the flesh. It is perfectly possible to simply score the skin, peel it back, then eat the flesh off the pit, but this is best done in the privacy of your own home, preferably in the bathtub, for not only does the juice go everywhere, but you also end up with fibers stuck in your teeth. It's worth it only if you don't want to waste a single sliver. A neater method is to leave the peel on and cut a lengthwise slice from either side of the pit. Turn each slice over so the skin is underneath. Cut the flesh grid-fashion, marking it into squares, but do not cut through the skin. Turn the skin inside out, and the squares of mango will stand proud and can easily be cut off. If you want slices rather than cubes, peel the mango, cut the flesh off the stone in big pieces, then slice it. Don't waste the flesh left on the pit; cut it off with a small, sharp knife.

mango platter

Take a tip from the Thais and arrange slices of mango, papaya, and pineapple in overlapping rows on a platter. Decorate with twists of lime and orange. This is the perfect dessert after a rich meal, and also makes a good addition to the buffet table. Set out toothpicks for spearing the pieces of fruit, which should be small enough to be eaten in one or two bites.

cooking

Although unripe mangoes are cooked as vegetables in some countries, the ripe fruit is best served fresh. Mango makes a good salsa when spiked with chilis, and chilled mango slices are great with curry.

mango ice cream

Many ice cream lovers believe this to be the finest flavor ever invented.

Serves 4

¾ cup sugar
3 egg yolks
1¼ cups milk
1 cup heavy cream
2 ripe mangoes (total weight about 1½ pounds)
2 tablespoons fresh orange juice
fresh mango slices, to serve (optional)

Have ready a saucepan and a heatproof bowl that will fit on top of it. Put half the sugar in the bowl and add the egg yolks. Whisk until pale and thick. Half fill the pan with water, making sure the level is below that of the bowl when placed on top. Heat the water to simmering point.

Heat the milk and cream in a separate pan. When the mixture boils, stir it into the egg yolks. Place the bowl over the simmering water and stir constantly until the mixture thickens to a custard that coats the back of a spoon. Leave the custard to cool, stirring it occasionally, then refrigerate.

Peel the mangoes. Slice the flesh off each pit and put it in a blender or food processor. Process until smooth, then add the remaining sugar and orange juice. Process briefly to mix.

Stir the mango purée into the chilled custard. Churn the mixture in an ice cream maker, or freeze in a shallow container, whisking in the ice crystals at regular intervals.

Let the ice cream soften a little before serving in scoops in ice cream cones, or in bowls, with slices of fresh mango.

tip

• The intensity of the mango flavor depends on the quality of the mangoes. If the flavor is a little weak, you can strengthen it by adding extra mango purée, made by cooking some dried mangoes in a little water until soft, then puréeing the mixture in a food processor. Alternatively, purée canned mangoes with a little of the can juices. Either way, stir the purée into the ice cream mixture before churning it.

white chocolate & mango terrine

Sumptuous, yet surprisingly simple, this dessert can be made ahead of time, so it is perfect for entertaining.

Serves 6

7 squares (7 ounces) white chocolate, broken up

3 tablespoons Amaretto liqueur or sherry

⅔ cup light cream

1¼ cups heavy cream

4 ounces almond macaroons or amaretti biscuits, crushed (1⅓ cups crumbs)

1 small mango, about 9 ounces

6 fresh Cape gooseberries, dipped in confectioners' sugar, to decorate

Sauce

1 large ripe mango, about 11 ounces

¾ cup orange juice

sugar (optional)

Line an 8½ x 4½-inch loaf pan with plastic wrap, allowing enough excess to fold over the pan after it has been filled.

Put the chocolate and liqueur or sherry in a heatproof bowl that will fit over a small saucepan. Add 2 tablespoons of the light cream. Half fill the pan with water and bring it to a boil, then remove it from the heat, place the bowl on top and set it aside until the chocolate melts. Stir in 2 more tablespoons of the light cream until the mixture is smooth. Let cool slightly.

Pour the rest of the light cream into a bowl, add the heavy cream, and whip until soft peaks form. Whisk in the melted chocolate mixture, then fold in the macaroon or amaretti crumbs and mix well. Spoon half the mixture into the prepared pan.

Peel the mango and slice the flesh off the pit. Cut it into thin strips. Lay these on top of the crumb mixture in the pan, then spoon the rest of the mixture on top. Fold back the plastic wrap and press lightly to ensure that the mixture is flat and fills the pan evenly. Freeze for 6 hours, or overnight.

Make the sauce. Peel the mango and slice the flesh off the pit. Put it in a blender or food processor with the orange juice. Blend until smooth. Scrape into a pitcher and stir in a little sugar, if needed.

Remove the terrine from the freezer, invert it on a platter and lift off the plastic wrap. As soon as it has softened sufficiently, slice it neatly. Spoon a little of the mango sauce onto each dessert plate and add a slice of the terrine. To use the Cape gooseberries, peel the husks back to expose the orange berries. Dip these in confectioners' sugar and place one on each plate.

melons

When you are hot and thirsty, what could be more refreshing than a slice of melon? Sweet and juicy, melon is perfect for every occasion, from a casual beach picnic to a formal dinner. Although we tend to see only a few varieties at the market, there are actually hundreds of different types. All have hard, inedible skins, juicy flesh, and numerous seeds, usually in the center of the fruit.

Melons have been cultivated for thousands of years in Asia and Africa, where desert peoples valued them as much for their thirst-quenching qualities as for their flavor. In ancient Egypt, melons were an important crop, and the fruit was relished by the ancient Greeks and Romans. In one of the world's earliest cookbooks, written by the Roman epicure, Apicius, there is a recipe for watermelon and "honeymelon" in a sweet wine sauce. His contemporary, the emperor Tiberius, is on record as being particularly fond of melon, and the fruit is depicted on murals that were discovered in the ruins of Herculaneum.

By the beginning of the 16th century, melons were being cultivated in Italy and the south of France. Christopher Columbus is generally credited with introducing them to America, although there is some evidence that a type of watermelon was grown by Native Americans in the Mississippi Valley long before his time. Today, melons are widely grown in both the northern and southern hemispheres, and supplies are in supermarkets all year round.

winter and summer melons

As with squash, to which they are related, there are both winter and summer melons. The former tend to be oval, with very thick skins. Most are sweet but some are less so, and are treated as vegetables in both Asia and Africa. Winter melons include the pale, green-skinned honeydew, with its delicate green flesh; the large yellow Casaba, with its creamy-white flesh; and the oval Crenshaw, whose flesh is salmon pink under skin that is green and yellow. Winter melons have little or no detectable aroma, unlike summer melons, which are often highly scented. Of the summer melons, cantaloupes are probably the best known. These are often described as netted melons, due to the characteristic markings on the skin. Cantaloupes are plump and round, like small beige beach balls. The

107

flesh (usually peach-orange but occasionally lime-green) is beautifully scented and sweet. Persian melons are similar, but larger. Two popular melons from Israel are the Galia and Ogen, the latter taking its name from the kibbutz on which it was developed.

Watermelons are in a category of their own. These thirst-quenching fruit are often very large, but smaller, seedless versions are now being produced which are much easier to carry home from the market and fit in the refrigerator. Most watermelons have pretty red flesh, but yellow-fleshed varieties are also grown.

nutrition

All melons have a high water content and are low in calories. The orange-fleshed varieties contain beta-carotene.

selection and storage

Choose firm, well shaped melons that are heavy for their size. Avoid any that have damaged skins, or look dented. The scar at the stem end should be smooth and clean; if it looks rough, or still has a bit of the stem attached, the fruit was probably picked too soon and may not have a good flavor. It is quite difficult to judge whether a winter melon is ripe, as the hard, odorless skins don't give much away. Canteloupes and other summer melons are a little easier, as the skin will give a little around the blossom end when the fruit is ready for eating, and the fruit will smell beautifully fragrant.

Buy whole fruit, if possible, but if you must buy a chunk, make sure the flesh is firm and crisp. Refrigerate it as soon as possible after purchase.

Other types of melon do not necessarily need to be kept in the refrigerator. If they are on the firm side, let them ripen at room temperature for a few days. If they are ripe, store them in a cool place, but only for a day or two. Alternatively, cut the flesh into balls or cubes, then chill them in a plastic tub. Close the tub tightly, or the fruit will impart its flavor to other foods.

preparation

Cut the melon in half and scoop out the seeds. The flesh can be cut off the skin and diced, or simply scooped out with a melon baller. Half a small melon makes a good appetizer. You can fill the center with port, pack it with lychees, or sprinkle the flesh with ground ginger. A popular way of presenting melons is as boats: Cut the fruit into wedges, then run a knife blade along the skin to loosen the flesh, but not remove it. Cut the flesh widthwise, and push alternate pieces to the left and right, so the melon wedge looks like an outrigger canoe. Melon is traditionally served with prosciutto.

watermelon fruit salad

Serve a fruit salad in a watermelon shell. Cut the watermelon in half, using a zig-zag pattern. Remove the seeds, then ball the flesh. Having done this, scoop out all the remaining flesh and use it to make a smoothie or a similar drink. Mix the watermelon balls with balls from green- and orange-fleshed melons, adding other types of fruit if you like. Dress lightly with freshly squeezed orange juice, and pile the fruit salad into the melon shell.

watermelon & white peach salad

This tasty savory salad is especially good with barbecued meats.

Serves 6

1 1½-pound piece of watermelon
2 white peaches, skinned and sliced
1 cucumber, halved and thinly sliced
1 cup diced feta cheese
fresh mint sprigs, to garnish

Dressing
2 tablespoons balsamic vinegar
1 teaspoon honey
6 tablespoons light olive oil
salt and pepper

Cut off the watermelon rind and slice the flesh into triangles, removing any seeds. Arrange around the edge of a large platter, with the points of the triangles toward the rim.

Make an inner circle from alternate slices of peach and cucumber. Repeat the circles, working inwards, until you have almost reached the center. Leave this empty.

To make the dressing, pour the vinegar into a pitcher and whisk in the honey until it has dissolved. Whisk in the olive oil, then add salt and pepper. Drizzle half the dressing over the salad, cover and let stand in a cool place for about an hour.

Just before serving, drizzle the remaining dressing over the salad. Heap the feta cubes in the center, and garnish with the mint sprigs.

melon pearls in rosewater dressing

An appetizer which works equally well as a dessert, this looks pretty and has a delicate flavor.

Serves 4–6

1 canteloupe
1 Ogen melon
½ cup sugar
¼ cup rosé or medium white wine
2 tablespoons rosewater
orange twists, to decorate

Cut both melons in half and scoop the seeds into a sieve set over a bowl. Press the seeds against the sieve to extract any juice, then discard them.

Working over the bowl, remove the flesh from each melon half with a small baller, at the same time allowing any juice to drop into the bowl. Stir in the sugar and wine, and chill for at least 2 hours.

Stir in the rosewater. Serve in dessert glasses or bowls, decorated with orange twists.

variations
• Use one type of melon and add grapes. Green grapes and Ogen melon look good, or for a dramatic color contrast, use cantaloupe balls with purple or black grapes.
• Serve the mixture in a vandyked melon shell (cut with a zig-zag edge), or serve individual portions in glass cocktail bowls set over crushed ice.
• If making this for children, omit the wine and use apple juice instead.

melons

bananas

They're such cheerful fruit, bananas. It's not just the bright yellow color, nor the easy zippability, nor even the way in which they have been portrayed in slapstick humor down the years. Perhaps it has something to do with the fact that they are highly nutritious and a great source of slow-release energy, so we feel good after eating one. The creamy, sweet flesh tastes delicious when ripe, and is so easy to digest, it is as suitable for a toddler as for a teenager inclined to skip meals. It gives the busy adult a boost, and is a nutritious choice for an elderly person who may be disinclined to bother with fruits that require more elaborate preparation.

The name banana comes from Africa, although the fruit is believed to have originated in Malaysia. It reached South America in the early 16th century, courtesy of a Spanish missionary, but did not catch on in the United States until 1876, when bananas wrapped in foil were sold for 10 cents a time at the Philadelphia Centennial Exposition.

Today, bananas are hugely popular all over the world, and it is rare to find a fruit bowl that does not contain them. There are several different types, including the tiny sugar bananas and the sweet-fleshed red bananas.

nutrition
A valuable carbohydrate food, the banana is an excellent source of potassium, and contains vitamins C and B6.

selection and storage
Bananas are picked while still green, and are stored in ripening rooms. The ends are the last parts to turn yellow, and the ones we find in our markets are often tipped with green. At this stage the flesh will be firm, with little flavor. Cook bananas like these, but don't eat them raw, or you may find them rather indigestible. Bananas that are golden all over are best for eating. As they ripen further, they will become speckled with brown and will finally develop dark patches on the skin. Even when that happens, it is worth checking the interior, which may still be

perfect, if a little soft. If you find blackened areas on the flesh, cut them out. Overripe bananas are perfect for making banana bread.

Buy bananas in bunches, rather than loose, and keep them that way until you eat them. They should be stored away from other fruit, preferably with air circulating around them, so those metal trees designed for the purpose really are worth having. Ripe bananas give off a gas that will help other fruits to ripen. You can keep bananas in the refrigerator, but the skins will blacken.

preparation

Peel bananas just before eating them, if possible. If you must slice them ahead of time, brush the flesh with lemon juice to prevent discoloration. Sliced bananas are often served with curries, and are good in fruit salads if added at the last moment. If there is fruit salad left over, fish the bananas out before putting the dessert in the refrigerator, or they will become mushy. Peeled bananas, brushed with lemon juice, can be frozen. Children love them cut in half and speared on popsicle sticks.

cooking

For grilling, baking, or frying, choose fruit that is yellow all over, but still reasonably firm. Bananas taste delicious with other fruits in a kebab. Brush with a mixture of fresh lime juice and clear honey before grilling. Use sliced bananas for tarts, puddings, or pies. Whole bananas can be barbecued in their skins. Put them on the grill after you have served the main course, when the embers are dying, and by the time you've finished your burgers or grilled vegetables, the skins will have blackened and softened, and the centers will be like banana custard. Some cooks make a slit through the banana skins and flesh, and insert chocolate chips before barbecuing.

flaming bananas

This is wickedly rich, but tastes delicious. It is traditionally served with vanilla ice cream, but have it with low-fat yogurt, if that makes you feel better!

Serves 6

1 stick butter
½ cup packed dark brown sugar
6 bananas, ripe but fairly firm
6 tablespoons dark rum
vanilla ice cream or yogurt, to serve

Melt the butter in a heavy-bottomed frying pan. Sprinkle in the sugar and stir over medium heat until it has dissolved.

Quickly peel the bananas and cut them in half lengthwise. Lay them in the syrupy mixture and cook them for 3–4 minutes, until tender, spooning the syrup over frequently.

Pour the rum into a metal soup ladle and heat it gently. Set it alight with a long-handled match, then pour it over the bananas. As soon as the flames die down, serve the bananas with vanilla ice cream or yogurt.

variations

• Marinate the banana halves in ½ cup fresh orange juice for 5 minutes. Having melted the butter, add the sugar and pour in the orange juice mixture. When the mixture bubbles, add the bananas and proceed as in the main recipe.
• Try substituting banana liqueur for part of the rum.
• If you don't like rum, try using brandy or bourbon.

caribbean banana & coconut loaf

Use very ripe bananas for this nutritious and delicious loaf. The coconut is an inspired addition and really complements the banana flavor.

Serves 8

2 cups all-purpose flour
2 teaspoons baking powder
pinch of salt
2 eggs
½ teaspoon pure vanilla extract
1 stick butter
½ cup sugar
3 tablespoons milk
3 tablespoons honey
3 very ripe bananas, mashed
⅔ cup dried coconut

Preheat the oven to 350°F. Line and grease a 2-pound loaf pan. Sift the flour with the baking powder and salt. Whisk the eggs with the vanilla extract until they are pale and thick.

Cream the butter with the sugar until light and fluffy. Fold in the flour mixture alternately with the whisked eggs, milk, and honey, then gently fold in the mashed bananas and coconut.

Spoon the mixture into the prepared pan and smooth the surface. Bake for 1 hour, or until a skewer inserted in the cake comes out clean. Cool in the pan for 5 minutes, then invert on a wire rack to cool completely.

banoffee tartlets

Slicing a whole banoffee pie can be a bit messy, so it makes sense to bake individual tarts.

Makes 18

12 ounces ready-made pie dough
Filling
1¼ sticks butter
¾ cup sugar
1 14-ounce can sweetened condensed milk
3 medium bananas
1 cup heavy cream

Preheat the oven to 400°F. Roll out the pie dough on a lightly floured surface, and line 18 2½-inch tartlet pans. Prick the bottom of each pastry case with a fork, then bake for 10 minutes. Leave to cool.

Make the filling by melting the butter with the sugar in a pan. Pour in the condensed milk. Cook, stirring constantly, until the mixture boils. Lower the heat and simmer, stirring all the time, until the mixture thickens and turns a rich golden brown. Remove from the heat.

Peel and slice two of the bananas and divide the slices among the tartlet pans. Pour the caramelized condensed milk over, filling each tartlet almost to the top and covering the bananas completely.

Cool, then chill for about 1 hour, during which time the filling will become thicker. Whip the cream and then pipe a swirl on each tartlet. Peel the remaining banana and slice it thinly. Cut each slice in half and arrange like butterfly wings on the cream. Serve at once.

papayas

Papayas are very nutritious fruits. If you eat half an average papaya, you'll have exceeded your daily requirement of vitamin C by 50 percent, and will have had a tenth of the amount of fiber recommended daily. If you have a sore throat, drink papaya juice to soothe it, and, if boxers are to be believed, rubbing papaya skin on the spot will alleviate bruising on the skin. This is thanks to an enzyme, papain, which, in addition to its other qualities, is an excellent tenderizer. Marinate meat in puréed papaya and it will become meltingly tender. Don't overdo it, though, or the meat will begin to disintegrate. Papain, like similar enzymes in fresh pineapple and kiwi fruit, negates the setting power of gelatin, so don't try to set the fruit in a jelly.

Papaya trees grow very quickly and yield a harvest after only one year. They are widely cultivated in tropical and subtropical zones. The oval fruit can be as heavy as 13 pounds, though the ones we see in supermarkets are considerably smaller than this. The skin is green at first, ripening to a rich yellow, although there is also a red-skinned variety. The highly perfumed flesh can be orange or salmon pink.

selection and storage
Papayas have soft skins and bruise easily, so must be handled with care. Do not buy fruit that is damaged or whose flesh looks scorched. Avoid very underripe, greenish yellow fruit, or fruit that is mushy. Look for papayas that are yellow all over and are beginning to soften. They will continue to ripen at room temperature. Ripe papayas can be kept in the refrigerator.

preparation and usage
Cut a papaya in half and you'll find a mass of grey-black seeds, looking rather like frogspawn, in the center. Scoop these out and throw them away, then serve the fruit as it is, with a little lime juice, or fill the center with a seafood salad. Slices of papaya look good on a mixed fruit platter. The fruit can also be

cut in wedges and served in the same way as melon or pineapple boats. Passion fruit goes very well with papaya, as does mango. The fruit aids digestion, so it is a good idea to serve it at the end of a meal.

Green papaya is served as a vegetable in some parts of the world.

papaya lamb kebabs

Papaya is a wonderful tenderizer and also adds flavor to meat that is marinated in the puréed fruit.

Serves 6

2 pounds boneless lamb, cut into 1-inch cubes
12 pearl onions
2 green peppers, cored, seeded, and cut into 1-inch squares
1 red pepper, cored, seeded, and cut into 1-inch squares
salt and pepper
fresh rosemary sprigs, to garnish

Marinade

1 small ripe papaya
juice of 2 limes
1 teaspoon garam masala
1 teaspoon ground coriander
2 tablespoons light olive oil
¼ cup plain yogurt

Make the marinade. Cut the papaya in half, spoon out and discard the black seeds, then scrape the flesh into a blender or food processor. Add the

remaining ingredients and purée until smooth. Spoon into a shallow glass dish, add the lamb cubes and mix well. Cover and marinate for 2–3 hours, stirring occasionally.

Put the unpeeled onions in a pan and pour over cold water. Bring to a boil and boil for 4 minutes, then drain. When cool enough to handle, slip off the skins.

Drain the lamb cubes, reserving the marinade. Thread them on 12 metal skewers, alternating them with the onions and pepper squares. Season well.

Preheat the broiler or fire up the barbecue. Cook the kebabs under or over medium heat for 15–20 minutes, turning them occasionally and basting them often with the marinade. Garnish with rosemary before serving.

tip

• Don't marinate the meat for any longer than 3 hours or it will become so tender it will start to disintegrate.

dates

The fruit of the date palm is one of the oldest foods known to man. High in natural sugars and a good source of fiber and minerals, it has provided sustenance for desert peoples in the Middle East and North Africa for thousands of years. Jars packed with dates have been found in the tombs of the pharaohs, and the fruit is known to have been cultivated by the Babylonians.

There are over 350 varieties of date, generally classified by moisture content. Soft dates are plump, voluptuous, and juicy. More fragile than other types, they must be harvested by hand. Semi-dry dates have firmer flesh and less moisture. These are harvested mechanically. Two well known examples of this type of date are the light brown Deglet Noor (the kind we buy in boxes at Christmas time) and the Zahidi. A third category—dry dates—have hard flesh and are so dry that they are also called "bread dates." These are seldom seen in our shops, but are a staple food for nomadic tribespeople of North Africa and the Middle East.

Perhaps the most delicious soft date is the Medjool. This large, wrinkled fruit was so highly regarded in its native Morocco that it was reserved for the royal family and their guests. Lesser mortals might never have got to taste it had it not been for a disease that began killing off the date palms in the 1920s. In an attempt to save the Medjool, immature palms were sent to the United States. They flourished in the arid conditions in Arizona and California, and the US is now the world's foremost producer of this type of date. Other important date producers include Israel, Iraq, and Tunisia.

In addition to the fresh fruit, dates are sold dried, either whole, compressed into blocks, or finely chopped and mixed with sugar. These products can be useful in baking, but cannot compare with the flavor and texture of a sweet, succulent fresh date.

selection and storage

When buying fresh dates, look for fruit that is plump and moist, but not too sticky. With the exception of Medjools, which have wrinkled skins, most dates should be smooth. Dates keep extremely well. Pack them in a sealed plastic tub so that they do not dry out too much, and store them in a cool place or the refrigerator.

preparation

If you need to peel dates, do so by squeezing the stem end. Eat the fruit whole, then discard the pits. Whole dates are delicious when filled with marzipan or cream cheese. Chopped dates are good in fruit salads, and can add a touch of sweetness to coleslaw or an orange and fresh leaf salad. Use a sharp knife or a pair of kitchen scissors for cutting, dipping the blades frequently in boiling water. Fresh or dried dates make a wonderful filling for baked apples or pears, and can also be used in place of sugar in cakes and bakes. For a quick dessert, snip fresh or dried dates into yogurt or mix them with mascarpone.

stuffed dates

Serve these after-dinner treats with cups of espresso coffee.

Serves 6

12 ripe fresh dates
4 ounces almond paste
whole almonds and small pieces of candied fruit,
     to decorate

Using a sharp knife, make a slit in each date, taking care not to cut them in half completely. Remove the pits.

Divide the almond paste into 12 pieces and shape to fit the date cavities. Reshape the dates, leaving a little of the almond paste visible.

Press a whole almond or a piece of candied fruit into the almond paste on each date. Arrange the stuffed dates on a plate, and serve with coffee.

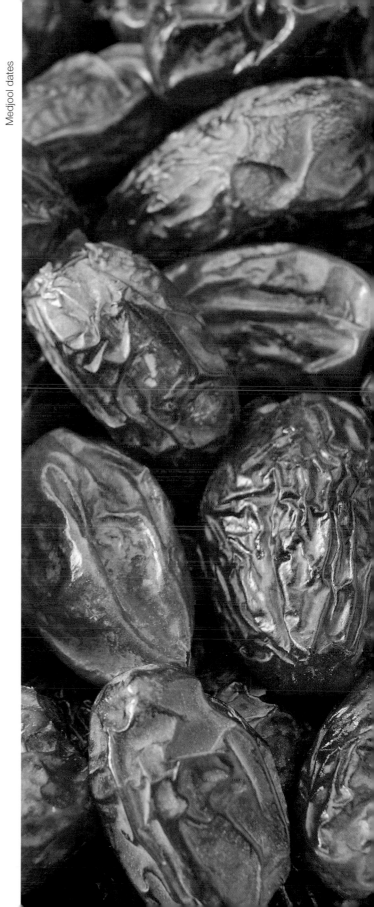

Medjool dates

guavas

Guavas haven't attained the designer status afforded to some exotic fruits, but they deserve our attention because they are very high in vitamin C and have an interesting flavor. They make an excellent contribution to fruit salads and other fruit desserts. Perhaps it is their odor that puts some people off. The fruit is highly perfumed, even musky; the scent of a single guava in a fruit bowl can pervade the whole house. Some people like the smell, but others find it rather overwhelming.

Guavas originated in South and Central America. South America remains a major producer, but the fruit is also grown in Asia, Africa, Australia, and the United States. Guavas can be round or oval. The shiny skin starts off green, but ripens to a pale yellow. There is also a pale pink variety, commonly known as the strawberry guava. Feijoas were originally thought to be related to guavas, and are sometimes called pineapple guavas, but they actually belong to a different family of fruit.

selection and storage

Look for guavas that are uniformly yellow. They should feel quite firm. Avoid any that are shrivelled or bruised, or whose skin is pitted. If there are fruits of various sizes, choose the larger ones, as small specimens tend to contain a higher proportion of seeds to flesh. Store guavas in a cool place, preferably the refrigerator, but wrap them carefully first or they will scent other foods.

preparation

Whether or not to eat the skin is a matter of choice. If the guavas are young and tender, you can eat them as they are, but if the skin is tough, you may prefer to cut the fruit in half, then scoop out the flesh, leaving the skin behind. A similar situation relates to the seeds: In young guavas these are soft and pale, and can be eaten, but if they are hard, you may prefer to discard them.

One of the most delicious ways of serving guavas is also the simplest; just slice them, spread out the slices on a platter, and sprinkle them lightly with sugar. Cover and leave overnight. Next morning, the sugar will have blended with the guava juice to make a delicious syrup. Serve the fruit with yogurt for breakfast. Raw ripe guava is good in a salad.

cooking

Guavas cease to smell if you cook them. Poach slices or chunks in a light syrup, but keep the heat low and watch them closely, as they rapidly cook down to a pulp. If this happens, blend the flesh in a food processor, press it through a sieve to remove the seeds, then use it as a filling for pancakes or as the basis of a creamy whipped dessert (fold the purée into whipped cream or a mixture of cream and ricotta cheese). The fruit has an affinity for apples and pears, and is a good addition to a pie or crisp.

iced guava cordial

Guava makes a good drink for both children and adults. Quarter 1 pound ripe guavas and put them in a pan. Add 2 cups water and bring to a boil. Simmer until the fruit is tender, then pour the contents of the pan through a sieve into a pitcher. Press the pulp through the sieve and stir it into the liquid, with 2 tablespoons of fresh lime juice. Add about 6 tablespoons of sugar to sweeten the mixture. To serve, pour equal amounts of cordial and soda water into tall glasses. Float a little grenadine on top, if you like.

guavas

119

lychees

Also known as litchis, these fascinating fruits came originally from southern China, where they were cultivated for more than 2,000 years before they came to the attention of the West. China is still an important producer, but today lychees are cultivated in tropical areas of many more countries, including India, the Philippines, Thailand, Madagascar, South Africa, Australia, and Israel.

The dark red skin of a lychee is thin, hard, scaly, and so brittle that you can break it with a thumbnail. Remove it, and you are in for a surprise. The fruit inside is a snow-white translucent globe, about the size of a large grape, with a dark brown pit in the center. Succulent, scented, and sweet, the flesh has a mysterious, exotic flavor. People who love lychees can easily get through a dozen or more at a sitting.

Lychees are very refreshing and are often served at the conclusion of a Chinese meal. Snapping the shells and biting the flesh straight off the pit is very satisfying, but the fruit can also be pitted and incorporated in a fruit salad. It goes well with orange, ginger, and passion fruit, and also makes a good partner for papaya.

Dried lychees are called lychee or litchi nuts. They are dark brown in color and look rather like prunes, with a slightly nutty flavor. Look for them in specialist food shops and serve them as after-dinner nibbles.

nutrition
Lychees are a source of vitamin C, and also yield some minerals (notably potassium) as well as dietary fiber.

selection and storage
Look for plump, unshrivelled fruit that feel heavy for their size. The shell-like skin should be dark pink or red, with no cracks. If the skin looks brown and dusty, the fruit will be past its prime and may taste bitter. Store lychees in a perforated plastic bag in the refrigerator. They will stay fresh for up to a week.

preparation and cooking
Simply crack the shells, lift them off, and eat the flesh off the pit. Lychees can also be halved, pitted,

and lightly poached in syrup. A squeeze of citrus juice brings out the flavor. Alternatively, use lychees in a stir-fry: They go well with pork or chicken. Add the peeled and pitted fruit right at the end, and toss the mixture over the heat until the lychees are just warm. Lychee pulp can be used to make sorbets, ice cream, or fruit drinks.

lychees with orange & ginger

A fresh-tasting dessert that looks as good as it tastes.

Serves 4

½ cup sugar
1 ¼ cups water
2 oranges
16 lychees, peeled and pitted
2 pieces drained preserved ginger, sliced
2 passion fruit

Heat the sugar and water in a small pan, stirring until the sugar has dissolved, then boil the syrup for 1 minute without stirring.

Pour the syrup into a serving bowl, and leave until cold. Peel the oranges and segment them, working over the bowl of syrup so that any juice is incorporated. Add the orange segments to the bowl with the lychees and ginger. Stir lightly to combine the ingredients.

Spoon into individual glass dishes. Cut the passion fruit in half and scoop the pulp over the fruit. Serve at once.

lychees

carambolas

Carambolas look like no other fruit on the stall. A glistening, golden yellow color, they are basically oval, but five ribs or "wings" give them a structured, almost artificial appearance, which is underlined by the waxy sheen on the thin skin. When you cut up a carambola, each slice forms a perfect star, which explains their alternative name—starfruit. The translucent flesh is juicy and has a slightly crunchy texture. Freshly picked ripe carambolas have a wonderful sweet-sour flavor, but this can be dulled when the fruit is transported vast distances from the countries where it is grown. Although the taste may be disappointing on occasion, there is ample compensation to be had from the contribution the beautifully shaped slices make to the appearance of a tropical fruit salad or savory dish.

The fruit is believed to have originated in Ceylon and the Moluccas, and is now widely cultivated in China, India, Australia, the Philippines, Tahiti, Hawaii, and the Caribbean. It goes under various names, including Star Apple, Country Gooseberry, Kamaranga, and the appropriate Five Corner. Carambolas were introduced into Florida in the late 19th century, and are now also cultivated in California, too. Arkin, Golden Star, and Hoku are among the bright yellow varieties, while Maha and Fwang Tung are paler. These are all sweet carambolas. There are also a number of sour varieties that can be used as a garnish, in much the same way as lemons or limes.
Carambolas are a good source of vitamin C, potassium, and fiber. They also contain beta-carotene, and are very low in calories.

selection and storage
Choose shiny, plump fruit that is brightly colored. The skin can be yellow, orange, or even the color of rich cream, but avoid carambolas whose skin is tinged with green. A brownish blush along the ridge tips is a sign that the fruit is fully ripe. Handle carambolas carefully, as they bruise easily. They can be stored at room temperature for a day or so, and will keep for up to a week in the refrigerator.

preparation

The waxy skin is edible, so there's no need to peel
the fruit, although some sources recommend paring
the skin lightly on the ridge tips. Carambolas are
usually sliced widthwise to show off the stellar
shape, but can be sliced lengthwise. The seeds,
which look rather like apple seeds, are edible, but
some people prefer to remove them. Star slices
from small fruit can be frozen and used instead of
ice cubes in drinks, but don't freeze the fruit and
attempt to thaw it again.

cooking

Ripe carambolas are usually eaten fresh, but can be
very lightly poached. The slices make a beautiful
garnish; serve them as they are, or sauté them
lightly in butter. Carambola slices also make a good
addition to a pork or chicken stir-fry. Add them right
at the end of cooking, and toss the mixture over the
heat until they are just warmed through. In some
countries, the green fruit is cooked in fish dishes
and curries.

tropical fruit salad

You can use any tropical fruit for this, but a
mixture of lychees, carambolas, kiwi fruit,
and mango slices looks pretty and tastes
delicious. For the liquid, use either a syrup
made by heating 1 cup sugar with $2\frac{1}{4}$ cups
water, or a mixture of apple juice and fresh
orange juice. If you use syrup, chill it
thoroughly before adding the fruit, and
sharpen the flavor with fresh lime juice. If
you like, add sliced bananas or passion
fruit pulp just before serving.

cape gooseberry

Inside their little lanterns, these golden berries are cushioned from bumps and bruises and keep extremely well. They have a wonderful, slightly tart flavor, and are delicious simply nibbled off the stem. It is as a decoration, however, that Cape gooseberries really score. Twist back the papery husk, half dip in melted chocolate, and you have a very pretty orange and dark brown berry capped by a lacy veil.

The fruit originated in the highlands of South America, where it still grows wild. A relative of the tomatillo, and of the garden plant known as the Chinese lantern, it was introduced to England in the late 18th century. Soon after, seeds were taken to the southernmost tip of South Africa, the Cape of Good Hope, where the plant flourished in sandy, well drained soil. It become so prolific and so popular that many people thought it was indigenous to the area, and gave it the name by which it is known today, Cape gooseberry.

nutrition
Cape gooseberries are an excellent source of vitamin A, and also yield vitamin C.

selection and storage
Cocooned inside their papery husks, the berries will stay sound for many weeks. When the husks turn pale and straw-colored, the berries are ripe and will have changed from green to a bright orange. Keep Cape gooseberries in the refrigerator in the original packaging, with the husks intact.

preparation and cooking
Cape gooseberries are wonderful just as they are, but also make good jam, as they are high in pectin. Poaching them lightly in syrup intensifies their flavor, and they are also delicious in tartlets or as a pie filling. They go very well with apples. If they are to be

cooked, take the berries out of the husks, then rinse in cold water to remove the sticky coating. Discard any berries that have blackened areas on the skin.

If you don't intend to cook the fruit, and wish to retain the husk as a decorative handle, slit it, then tear it into "petals," leaving these attached to the berry. Twist each of these strips back, then pinch them together gently just above the fruit. Wipe the berries clean with dampened kitchen paper towels. They can then be coated in confectioners' sugar or fondant, or half dipped in chocolate.

Dried Cape gooseberries resemble raisins, and can be used in much the same way.

cape gooseberries dipped in chocolate

These make a pretty decoration for desserts—and taste delicious.

Makes 24

3 ounces dark chocolate
24 Cape gooseberries (total weight about 3½ ounces)

Bring a small pan of water to a boil. Break up the chocolate and put it in a small heatproof bowl that will fit over the pan without touching the water.

Remove the pan from the heat, place the bowl of chocolate on top and leave for 3–4 minutes, until the chocolate has melted.

Line a large baking sheet with parchment or wax paper. Slit the husk on each Cape gooseberry and

cape gooseberries

tear it into "petals", keeping it attached to the berry. Twist each of these strips back, pinching them together gently behind the golden berry so that it is completely free and the husk forms a decorative holder. Wipe the berries with kitchen paper towels.

Stir the melted chocolate until smooth. Lift each Cape gooseberry by the husk, holding just above the berry, and half dip it in the melted chocolate so that the berry remains partially visible. Place on the paper-lined baking sheet and leave in a cool place (not the refrigerator) to set. Serve as sweetmeats or use to decorate cakes or desserts.

cape gooseberry

passion fruit

Also known as the granadilla or maracuya, the passion fruit is a native of South and Central America, but is widely grown in Australia, South Africa, Kenya, and Zimbabwe, as well as the Mediterranean basin. Small, purplish-brown and wrinkly, it is not very impressive to look at, and certainly doesn't live up to the promise of its name.

Despite its unpromising appearance, the oval or round fruit is absolutely delicious. The leathery skin encases a mass of small black seeds surrounded by sweet, fragrant yellow pulp. Everything inside the skin is edible, although to improve the appearance of sorbets and similar dishes, the pulp is often sieved to remove the seeds.

Passion fruit has a wonderful aroma, suggestive of citrus but more highly scented. The pulp of a ripe fruit is sweet, and a small amount will contribute considerably to the flavor of a mixed fruit dessert such as a fruit salad. Passion fruit is an excellent source of fiber, and contains vitamins A and C.

In addition to the common purple passion fruit, look out for a larger yellow variety from Brazil. Another member of the same family is the Curuba. This South American native originated in the high Andes, but is now also cultivated in New Zealand. Flat and oval in shape, it has thick greenish-yellow skin. When sliced down the middle, the two halves look like boats filled with jelly-like seeded pulp. This resembles that of a regular passion fruit, but is not as sweet.

selection and storage

As a passion fruit ripens, the skin hardens and wrinkles and the pulp inside becomes steadily sweeter. Wrinkles are therefore desirable, but the fruit should not be so corrugated that it has begun to dry out. If it has a dusty and dented appearance, leave it alone. Store passion fruit in a cool place, preferably in a perforated plastic bag in the refrigerator.

preparation

This is simplicity itself. Just cut the fruit in half and use a teaspoon to scrape out the pulp. If you need to remove the seeds, spoon the pulp into a sieve set over a bowl and use the back of the spoon to press it through. Use the pulp as a topping for a cheesecake, or mix it with mascarpone and cream as a filling for brandy snaps or cream horns. Passion fruit makes wonderful ice cream, especially if you mix it with persimmon. It also has an affinity for peaches and nectarines, and tastes good spooned over melon balls. Sieved passion fruit pulp can be used to make a refreshing drink. One of the finest uses for passion fruit, however, is to make a curd or fruit cheese.

passion fruit curd

Delicious on toast, stirred into mascarpone, or just spooned from the jar.

Makes about 1 cup

¾ stick butter
pulp of 3 medium, or 2 large, passion fruit
juice of ½ lemon
2 egg yolks
½ cup sugar

Melt the butter in a heatproof bowl over simmering water. Off the heat, stir in the passion fruit pulp (seeds and all), lemon juice, egg yolks, and sugar. Replace over the simmering water and cook, stirring constantly until the mixture thickens. Pour into a sterilized jar and seal. When cold, keep in the refrigerator.

brandy snaps with passion fruit mascarpone

Brandy snaps seldom contain brandy but are lacy ginger cookies that are curled around the handle of a wooden spoon. They turn crisp as they cool and taste wonderful with a creamy mascarpone and passion fruit filling.

Serves 4–6

½ stick butter, plus extra for greasing
8 tablespoons all-purpose flour
1 teaspoon ground ginger
½ teaspoon ground allspice (optional)
¼ cup sugar
2 tablespoons golden syrup
1 teaspoon grated lemon zest
1 tablespoon lemon juice
1 cup mascarpone
2 passion fruit
2 tablespoons confectioners' sugar
⅔ cup heavy cream

Preheat the oven to 350°F. Grease two baking sheets and the handle of a wooden spoon with melted butter.

Sift the flour together with the ground ginger. Add the allspice, if using. Melt the measured butter with the sugar and syrup. Remove from the heat and gradually stir in the flour mixture, then add the lemon zest and juice. Beat the mixture vigorously until it is all thoroughly combined, with no lumps remaining.

Unless you have a team of willing helpers, bake the brandy snaps in small batches, so you can roll the hot cookies before they cool. Drop teaspoons of the batter on one of the greased baking sheets, leaving plenty of room—at least 3 inches—for spreading. Bake the cookies for 6–8 minutes, until they have spread and turned a rich golden brown.

Remove the baking sheet from the oven and place it on top of the warmer. Immediately place the handle of the wooden spoon on one end of one of the brandy snaps. Use a flat-bladed knife to flip the cookies over the spoon handle, then quickly roll the cookie round the handle to make a tube. Ease the brandy snap off the handle and quickly repeat the process with the remaining cookies. If they become too brittle to roll, warm them in the oven until they become pliable again.

Make more batches of brandy snaps in the same way, using the baking sheets alternately and greasing them each time. While the last batch is baking, make the filling. Put the mascarpone in a bowl. Cut the passion fruit in half and scoop the pulp into the bowl. Beat in the confectioners' sugar.

In a separate bowl, whip the cream to soft peaks, then fold it into the mascarpone mixture. Fill a forcing bag with the mixture and pipe a little into either end of each cold brandy snap. Don't attempt to fill the brandy snaps completely. Serve at once.

tip
• Don't fill the brandy snaps too far ahead of serving, or they will go soggy.

passion fruit pavlova

The perfect pavlova should be crisp on the outside and like marshmallow on the inside.

Serves 8

cornstarch, for dusting
6 egg whites
a pinch of salt
1½ cups sugar
1 tablespoon white wine vinegar
1¼ cups heavy cream, whipped
3 passion fruit

Preheat the oven to 300°F. Grease a 9½-inch springform cake pan, and line the bottom with parchment or wax paper and dust with cornstarch.

Whisk the egg whites with the salt until stiff peaks form. Setting aside 2 tablespoons of the sugar, gradually whisk in the remaining sugar. Add the vinegar with the final couple of spoonfuls of sugar. The mixture should be very stiff.

Scrape the mixture into the prepared pan and level the surface. Bake for 30 minutes, then lower the oven temperature to 250°F and bake for 1½ hours more. Switch off the oven and leave the pavlova inside until it has cooled completely.

Whip the cream with the reserved sugar. Cut the passion fruit in half and scoop the pulp into a bowl.

Transfer the pavlova to a plate—take care, as it will be very delicate. Carefully spread half the whipped cream over the surface, put the rest in a piping bag fitted with a star nozzle and decorate the rim of the pavlova. Spoon the passion fruit pulp over the center. Serve at once.

grapes

Few fruits give as much pleasure as grapes. When fresh they are deliciously refreshing; when dried (as currants and raisins), they provide concentrated sweetness and flavor. The juice is nourishing and thirst-quenching, and when transformed into good wine, it is one of the wonders of the world. The ancient Egyptians certainly thought so. Wine played a significant role in their rituals, and hieroglyphs found in tombs implore the gods to provide the deceased with wine "and every sweet thing" in the netherworld.

Grapes have been cultivated for thousands of years. Most of the grapes we enjoy today are derived from a wild variety that originated in the region close to the Caspian Sea, but wild grapes—or grape-like fruit—are known to have existed in other countries too. When the Italian explorer Giovanni da Verrazano reached North Carolina in 1524, he reported that many vines grew naturally there. These were scuppernongs, which are still cultivated today. Concord is a native American grape developed from wild fruit.

Today, we tend to differentiate between grapes grown principally for eating (table grapes) and grapes grown for wine-making. Italy is the world's premier producer of table grapes, followed by Chile and the United States. There are dozens of different varieties in colors that range from palest green and yellow, through red to deep blue-black or purple. Size and shape varies, too. Some are small and pearl-shaped, others oval. At least one variety, the Red Globe, has fruits as large as plums. Some of the seeded grapes have wonderful flavor, but tend to be rejected by shoppers seeking out the more convenient seedless varieties such as the green Thompson Seedless, the red Flame and Ruby Seedless, and the blue-black Beauty Seedless.

nutrition
High in natural sugars, grapes also yield vitamin C and small amounts of vitamin A, plus potassium.

selection and storage
Look for bright, clean grapes that adhere firmly to their stems. All the grapes on a bunch should be the

preparation and cooking

Most grapes need no preparation other than washing, but if you have chosen a thick-skinned variety, you may wish to peel them. Dip the grapes very briefly in boiling water, then remove the skins with a sharp knife. To seed grapes, cut them in half and ease the seeds out with a knife tip. For fruit salads and similar desserts, use a seedless variety, such as Thompson seedless.

Grapes are the classic accompaniment for the cheeseboard, and make a decorative centerpiece for the dinner table if presented on a footed bowl. Alternatively, serve grapes and cheese together in a mold. Glistening in a layer of grape juice jelly, dark grapes make a beautiful contrast to the cream cheese mousse beneath.

Grapes work well in a savory context. Sole Veronique is a famous fish dish that has grapes in the sauce, but grapes also go well with chicken, game birds, and pork. Finally, don't forget grape jelly. To make this popular preserve, use a variety such as Concord, which is highly flavored and slightly tart.

frosted grapes

These look very pretty as a decoration. Lightly whisk an egg white. If using a whole bunch of grapes, brush each fruit lightly with the egg white, then sprinkle all the grapes evenly with sugar.
• Miniature bunches of grapes can be dipped in egg white, then in the sugar. Either way, shake off the excess sugar, then dry on a wire rack.

same size. Ripe green grapes will have a subtle amber sheen, while red or blue-black grapes should be evenly colored all over. Transport grapes carefully, as they bruise easily. When you get home, pick off any damaged or split grapes and store the rest of the bunch in a perforated plastic bag in the refrigerator. Grapes are often served chilled, but it is better to let them come to room temperature so that their full flavor can be appreciated. Wash and dry them just before serving.

grape and cheese mousse

Black grapes, white cheese mousse—the combination not only looks dramatic, it tastes delicious with crackers and makes a welcome change from more conventional meal endings.

Serves 6

2½ cups red grape juice
5 teaspoons powdered gelatin
6 ounces seedless black grapes

Cheese Mousse
⅔ cup white grape juice
2 teaspoons powdered gelatin
1 7-ounce package cream cheese
¾ cup light cream

green grapes

Put 2 tablespoons of the red grape juice in a heatproof bowl and sprinkle the gelatin on top. Leave until the mixture is spongy, then place the bowl over simmering water until the gelatin has dissolved. Warm the remaining grape juice slightly, pour it into a large pitcher, and stir in the dissolved gelatin. Pour enough of the mixture into a 5-cup mold to make a layer about ¾ inch thick. Put the mold in the refrigerator until the mixture has set. Leave the remaining grape juice and gelatin mixture at room temperature.

Arrange the grapes over the layer of grape juice jelly in the mold, then carefully pour on just enough of the liquid gelatin mixture to cover them. Chill until set solid, then carefully pour the remaining gelatin mixture into the mold and chill for 3–4 hours until firm.

For the cheese mousse layer, spoon 2 teaspoons of the white grape juice into a cup and sprinkle the gelatin on top. Leave until spongy, then melt over simmering water. Warm the remaining white grape juice and stir in the dissolved gelatin. Leave to cool, but do not allow the mixture to set.

Beat the cream cheese until softened, then beat in the cream. Gradually add the white grape juice mixture, beating constantly. Pour the mixture into a bowl and chill until mousse-like and beginning to set.

Gently spoon the cheese mixture over the set jelly in the mold, filling it to the top. Level the surface and chill for at least 4 hours, preferably overnight, until the cheese layer has set. Carefully unmold the ring onto a serving plate. Serve in slices, with savory crackers.

grapes

tamarillos

This brightly colored fruit is a distant relative of the tomato but is shaped like a plum, which perhaps explains how it acquired two of its alternative names, tree tomato and Java plum. Tamarillos are widely grown throughout South America. New Zealand is a major producer, and the fruit is also cultivated in Australia, Southern Asia, and California.

There are several colors of tamarillo, from bright red and golden yellow to orange and even purple. The dark purple and red fruits have dark red flesh that is almost black in some cases, the yellow and orange varieties have yellow flesh. The seeds are edible.

nutrition
Tamarillos contain plenty of vitamin C, some vitamin A, and calcium. They are a good source of fiber.

selection and storage
Choose bright, smooth-skinned fruit that looks plump and fresh. When ripe, tamarillos smell fragrant and are just soft to the touch. Avoid any that have surface blemishes or feel soft. Tamarillos will ripen at room temperature; when they are ripe store them in a perforated bag in the refrigerator. They will keep for several weeks.

preparation and cooking
The bitter skin of the tamarillo is best removed. It comes off easily if the fruit is submerged briefly in boiling water. Red tamarillos are more tart than yellow ones, but all varieties need to be sweetened before being eaten unless you are serving them as a vegetable (they make excellent salsa). Be careful when cutting the fruit—the juice stains. Tamarillos are usually cooked, but some of the newer cultivars are so sweet that they can be eaten raw. (Look for Goldmine, Ecuadorian Orange, and Rothamer.)

One way of eating tart tamarillos is to cut them in half lengthwise, sprinkle the flesh with sugar, and leave them to stand overnight. Then scoop the flesh out of the skins. Alternatively, peel and slice the fruit, then cook it with plenty of sugar. Tamarillos taste wonderful in tarts, pies, and shortcakes. They are very juicy, so if you use them in a tart or pie, sprinkle the pastry with bread crumbs to soak up some of the liquid. Tamarillos make good jam and chutney.

tamarillo & coconut shortcake

This is an unusual recipe, but works wonderfully well. The coconut and sugar mixture soaks up the tamarillo juice to make a deliciously moist, sweet filling.

Serves 8

5 tamarillos, about 14oz
2½ cups all-purpose flour
1 tablespoon baking powder
½ teaspoon salt
1¼ sticks butter
1 cup caster sugar
½ cup flaked coconut
2 eggs, beaten
crème fraiche or yogurt, to serve

Preheat the oven to 400°F. Line and grease a 9-inch cake pan. Skin the tamarillos by placing them in a pan of boiling water for 30 seconds. Drain them, cut off the green tops and pieces of stalk, then peel off the skins. Let them cool, then slice them thickly.

Mix the flour, baking powder, and salt in a bowl. Cut in the butter, then rub it in until the mixture resembles bread crumbs. Stir in ¼ cup of the sugar. Tip the rest of the sugar into a bowl and stir in the coconut.

Add the beaten eggs to the dry ingredients to make a soft dough. Roll out half the dough on a lightly floured surface to fit the cake pan. Place it in the pan, then sprinkle half the coconut mixture evenly over the dough. Arrange the tamarillo slices in a single layer on top. Sprinkle the remaining coconut mixture over.

Roll out the remaining dough to a 9-inch round and place it carefully on top of the coconut-covered tamarillos. Press it down gently.

Bake the shortcake for 30 minutes until pale golden brown. Leave to stand in the pan for 10 minutes, then run a knife between the cake and the pan. Invert a plate over the shortcake, then turn the pan and the plate over together. Carefully lift off the pan. Serve the shortcake warm, with crème fraiche or yogurt.

variation

• Halved gooseberries or pieces of rhubarb can be used instead of tamarillos. Alternatively, use sliced apricots or nectarines, reducing the amount of sugar added with the coconut.

tamarillos

tamarillos

smoothies

Fresh fruit juices make wonderfully healthy and refreshing drinks, whether you serve them solo, with mixers, or as the basis for more elaborate beverages.

drink	ingredients	method
banana & honey smoothie This silky smooth drink is very satisfying. **Serves 4**	2 medium bananas 3 cups milk 3 tablespoons honey 1 scoop vanilla ice cream	Peel the bananas and chop them. Put them in a blender or food processor with 1 cup of the milk and blend until smooth. Add the remaining milk, with the honey and ice cream, and blend again. Pour into four tall glasses and serve.
cherryberry nectar Serve this with ice cubes into which you have set a whole pitted cherry. **Serves 4**	1 cup pitted sweet cherries 1 cup strawberries, hulled 2 cups milk ⅔ cup plain yogurt 3 tablespoons honey cherry-filled ice cubes	Put the cherries and strawberries in a blender or food processor and add the milk, yogurt, and honey. Blend until smooth. Pour into tall glasses and add a couple of cherry-filled ice cubes to each.

drink	ingredients	method
## lemon & lime cooler When the weather is hot, nothing cools you more quickly. **Serves 4**	½ cup sugar ¾ cup water 1 lemon 2 limes 8 ice cubes 2½ cups sparkling mineral water 4 frozen carambola slices, to decorate	Mix the sugar and water in a small pan. Heat gently, stirring until the sugar dissolves, then boil without stirring for 2 minutes. Cool, then pour into a 1-quart pitcher. Squeeze the lemon and limes, and strain the juice into the pitcher. Add the ice cubes, top up with the sparkling mineral water, and serve in tall glasses. Add a frozen carambola slice to each glass.
## pear & maple soother This is aptly named. Try it next time your throat feels a little raw. **Serves 4–6**	4 ripe pears 2½ cups chilled apple juice 1 tablespoon maple syrup, or to taste	Peel and core the pears. Chop them roughly and put them in a blender or food processor with the apple juice. Add the maple syrup. Blend until smooth. Serve in tall glasses. **tip •** Don't make this ahead of time, as it darkens on standing. This doesn't alter the flavor, but the drink doesn't look quite as appetizing as when freshly made.
## apple-nut energizer If the apple juice is reasonably sweet, you shouldn't need to add any honey or sugar to this rich drink. Taste it and see. **Serves 4**	¾ cup cashew nuts ¾ cup blanched almonds ½ cup water 1½ cups apple juice **tip •** This makes a very acceptable dairy-free ice cream. Sweeten it a little (chilling dulls the flavor slightly), then chill the mixture rapidly in an ice cream maker.	Grind the cashews and almonds very finely in a nut mill or coffee grinder. Tip into a tall pitcher and gradually stir in the water to make a smooth paste. Gradually whisk in the apple juice. The drink will still be quite grainy. If you have a hand blender, use it to improve the texture of the drink. Serve very cold, in small glasses.

smoothies

drink	ingredients	method

melon & mango dreams

One sip and you'll be dreaming of tropical beaches.

Serves 6

1 small cantaloupe
1 mango
juice of 4 oranges

Cut the cantaloupe in half. Scoop out and discard the seeds from both halves, then scoop the flesh into a blender or food processor.

Peel the mango and cut the flesh off the pit in large pieces. Add to the blender or processor. Process until smooth, then pour into a tall pitcher and stir in the fresh orange juice. Chill thoroughly. Serve in tall glasses.

tip • You can add sugar to this drink, but if the fruit is ripe enough, you shouldn't need to. For a special treat, add a scoop of orange sorbet or mango ice cream.

cranberry cup

Refreshing and very good for you, this drink has a lovely rich color.

Serves 6

3 cups fresh cranberries
1½ cups water
1 cinnamon stick
6 tablespoons sugar, or to taste
2 tablespoons lemon juice
juice of 4 oranges
seltzer water or lemonade to serve

Put the cranberries in a heavy-bottomed pan and pour in the water. Add the cinnamon stick. Bring to simmering point. Cook the cranberries for 5 minutes, or until they are very soft and have begun to pop. Press them through a sieve into a large pitcher. Stir in the sugar and lemon juice, and set aside.

When the cranberry mixture is cold, stir in the orange juice.

Chill well before serving in tall glasses, topped up with seltzer or lemonade.

smoothies

139

TYPE OF FRUIT	ENERGY (k)	ENERGY (Kcal)	PROTEIN (g)	FIBER TOTAL DIETARY (g)	VITAMIN A (IU)	THIAMIN VITAMIN B1 (mg)	RIBOFLAVIN VITAMIN B2 (mg)
Apples	247	59	0.19	2.7	53	0.02	0.01
Apricots	201	48	1.4	2.4	2612	0.03	0.04
Bananas	385	92	1.03	2.4	81	0.05	0.1
Blackberries	218	52	0.72	5.3	165	0.03	0.04
Blueberries	234	56	0.67	2.7	100	0.05	0.05
Cape gooseberries	222	53	1.9	*	720	0.11	0.04
Carambolas	138	33	0.54	2.7	493	0.03	0.03
Cherries, sour, red	209	50	1	1.6	1283	0.03	0.04
Cherries, sweet	301	72	1.2	2.3	214	0.05	0.06
Cranberries	205	49	0.39	4.2	46	0.03	0.02
Currants, black	264	63	1.4	*	230	0.05	0.05
Currants, red and white	234	56	1.4	4.3	120	0.04	0.05
Dates	1151	275	1.97	7.5	50	0.09	0.1
Figs	310	74	0.75	3.3	142	0.06	0.05
Gooseberries	184	44	0.88	4.3	290	0.04	0.03
Grapefruit, pink and red	126	30	0.55	*	259	0.03	0.02
Grapefruit, white	138	33	0.69	1.1	10	0.04	0.02
Grapes, red or green	297	71	0.66	1	73	0.09	0.06
Guavas, common	213	51	0.82	5.4	792	0.05	0.05
Guavas, strawberry	289	69	0.58	5.4	90	0.03	0.03
Kiwi fruit	255	61	0.99	3.4	175	0.02	0.05
Kumquats	264	63	0.9	6.6	302	0.08	0.1
Lemons	84	20	1.2	4.7	30	0.05	0.04
Limes	126	30	0.7	2.8	10	0.03	0.02
Mangoes	272	65	0.51	1.8	3894	0.06	0.06
Melons, cantaloupe	146	35	0.88	0.8	3224	0.04	0.02
Melons, casaba	109	26	0.9	0.8	30	0.06	0.02
Melons, honeydew	146	35	0.46	0.6	40	0.08	0.02
Nectarines	205	49	0.94	1.6	736	0.02	0.04
Oranges	197	47	0.94	2.4	205	0.09	0.04
Papayas	163	39	0.61	1.8	284	0.03	0.03
Passion fruit	406	97	2.2	10.4	700	·0	0.13
Peaches	180	43	0.7	2	535	0.02	0.04
Pears	59	59	0.39	2.4	20	0.02	0.04
Persimmons, Japanese	293	70	0.58	3.6	2167	0.03	0.02
Persimmons, native	531	127	0.8	*	*	*	*
Pineapples	205	49	0.39	1.2	23	0.09	0.04
Plums	230	55	0.79	1.5	323	0.04	0.1
Pomegranates	285	68	0.95	0.6	0	0.03	0.03
Quince	238	57	0.4	1.9	40	0.02	0.03
Raspberries	205	49	0.91	6.8	130	0.03	0.09
Rhubarb	88	21	0.9	1.8	100	0.02	0.03
Strawberries	126	30	0.61	2.3	27	0.02	0.07
Tamarillos	251	60	0.72	*	3	0.01	0.01
Tangerines and mandarins	184	44	0.63	2.3	920	0.11	0.02

Value given is per 100 grams of edible portion.

* indicates data is not available.

Source: US Department of Agricultural Research Service.

NIACIN VITAMIN B3 (mg)	VITAMIN B6 (mg)	VITAMIN C (mg)	VITAMIN E (mg)	CALCIUM (mg)	PHOSPHOROUS (mg)	IRON (mg)	POTASSIUM (mg)	MAGNESIUM (mg)	ZINC (mg)	MANGANESE (mg)	SELENIUM (mcg)
0.08	0.05	5.7	0.32	7	7	0.18	115	5	0.04	0.05	0.3
0.6	0.05	10	0.89	14	19	0.54	296	8	0.26	0.08	0.4
0.54	0.58	9.1	0.27	6	20	0.31	396	29	0.16	0.15	1.1
0.4	0.06	21	0.71	32	21	0.57	196	20	0.27	1.29	0.6
0.36	0.04	13	1	6	10	0.17	89	5	0.11	0.28	0.6
2.8	*	11	*	9	40	1	*	*	*	*	*
0.41	0.1	21.2	0.37	4	16	0.26	163	9	0.11	0.08	0.6
0.4	0.04	10	0.13	16	15	0.32	173	9	0.1	0.11	0.4
0.4	0.04	7	0.13	15	19	0.39	224	11	0.06	0.09	0.6
0.1	0.07	13.5	0.1	7	9	0.2	71	5	0.13	0.16	0.6
0.3	0.07	181	0.1	55	59	1.54	322	24	0.27	0.26	*
0.1	0.07	41	0.1	33	44	1	275	13	0.23	0.19	0.6
2.2	0.19	*	0.1	32	40	1.15	652	35	0.29	0.3	1.9
0.4	0.11	2	0.89	35	14	0.37	232	17	0.15	0.13	0.6
0.3	0.08	27.7	0.37	25	27	0.31	198	10	0.12	0.14	0.6
0.19	0.04	38.1	*	11	9	0.12	120	8	0.07	0.01	*
0.27	0.04	33.3	0.25	12	8	0.06	148	9	0.07	0.01	1.4
0.3	0.11	10.8	0.7	11	13	0.26	185	6	0.05	0.06	0.2
1.2	0.14	183.5	1.12	20	25	0.31	284	10	0.23	0.14	0.6
0.6	*	37	*	21	27	0.22	292	17	*	*	*
0.5	0.09	98	1.12	26	40	0.41	332	30	0.17	*	0.6
0.5	0.06	37.4	0.24	44	19	0.39	195	13	0.08	0.09	0.6
0.2	0.11	77	*	61	15	0.7	145	12	0.1	*	*
0.2	0.04	29.1	0.24	33	18	0.6	102	6	0.11	0.01	0.4
0.58	0.13	27.7	1.12	10	11	0.13	156	9	0.04	0.03	0.8
0.57	0.12	42.2	0.15	11	17	0.21	309	11	0.16	0.05	0.4
0.4	0.12	16	0.15	5	7	0.4	210	8	0.16	*	0.3
0.6	0.06	24.8	0.15	6	10	0.07	271	7	0.07	0.02	0.4
0.99	0.03	5.4	0.89	5	16	0.15	212	8	0.09	0.04	0.4
0.28	0.06	53.2	0.24	40	14	0.1	181	10	0.07	0.03	0.5
0.34	0.02	61.8	1.12	24	5	0.1	257	10	0.07	0.01	0.6
1.5	0.1	30	1.12	12	68	1.6	348	29	0.1	*	0.6
0.99	0.02	6.6	0.7	5	12	0.11	197	7	0.14	0.05	0.4
0.1	0.02	4	0.5	11	11	0.25	125	6	0.12	0.08	1
0.1	0.1	7.5	0.59	8	17	0.15	161	9	0.11	0.36	0.6
*	*	66	*	27	26	2.5	310	*	*	*	*
0.42	0.09	15.4	0.1	7	7	0.37	113	14	0.08	1.65	0.6
0.5	0.08	9.5	0.6	4	10	0.1	172	7	0.1	0.05	0.5
0.3	0.11	6.1	0.55	3	8	0.3	259	3	0.12	*	0.6
0.2	0.04	15	0.55	11	17	0.7	197	8	0.04	*	0.6
0.9	0.06	25	0.45	22	12	0.57	152	18	0.46	1.01	0.6
0.3	0.02	8	0.2	86	14	0.22	288	12	0.1	0.2	1.1
0.23	0.06	56.7	0.14	14	19	0.38	166	10	0.13	0.29	0.7
0.26	0.04	14.3	*	19	17	0.19	79	15	*	*	*
0.16	0.07	30.8	0.24	14	10	0.1	157	12	0.24	0.03	0.5

notes

use of raw eggs

Some recipes in this book contain uncooked eggs. These recipes are not suitable for very young children, the elderly, pregnant or nursing women, and people with compromised immune systems.

measurements

All the recipes in this book have been tested using standard tablespoons, liquid measures, and cups. This should make for absolute accuracy, but unfortunately not all sets of measuring cups on sale meet standard specifications. Much depends, too, on how you measure. In many recipes, minor inconsistencies will not matter, but for cakes and pastries, measuring accurately is vital.

For this book, metal measuring cups were used in the following sizes: 1 cup (8 ounces); ½ cup; ⅓ cup, and ¼ cup. Unless the recipe specifies that an ingredient is to be packed into a cup measure, ingredients like flour, cornstarch, and sugar have been lightly scooped into measures, then leveled off with the back of a knife. A cup of flour measured in this way is equivalent to 4 ounces; a cup of white sugar equals 8 ounces.

Follow methods closely. If, for instance, a cake recipe requires you to fold in an ingredient such as whisked eggs, use a gentle figure-of-eight action. If you beat the eggs in, you will knock out the air introduced to the eggs by whisking, and the cake will not be as light as it should be.

Where a recipe calls for 2 apples or 3 pears, use fruit of average size, if size is crucial, a weight will be given. All eggs used are large.

Ovens can vary. If you have a fan in your oven, circulating the hot air, you may need to use a slightly lower temperature than that recommended in the recipe. Check your owner's manual or seek advice from the manufacturer of your appliance.